ANTONELLA BONOMI

AN AFRICAN SONG

My life amongst the Maasai

The stories and anecdotes in this book are based solely on the author's recollection of events, circumstances, and occurrences. Some names, locations, settings, and identifying characteristics have been changed or modified to protect the privacy of the individuals and families depicted. While written as accurately as possible, dialogue has often been re-created from memory and not always from documented sources and therefore may differ from how other people remember it.

To Luca,
My eternal life companion,
with gratitude and appreciation
for having taught me to love
and understand Africa.
To my children, Lucrezia, Jacopo and Lorenzo,
with immense love,
and the hope that they too may one day write
their African diary.
And to those who are no longer with us but watch over them.
Always.

"Until one is committed, there is hesitancy, the chance to draw back, always ineffectiveness. Concerning all acts of initiative (and creation), there is one elementary truth, the ignorance of which kills countless ideas and splendid plans: that the moment one definitely commits oneself, then providence moves too. A whole stream of events issues from the decision, raising in one's favor all manner of unforeseen incidents, meetings, and material assistance, which no man could have dreamt would have come his way."

William H. Murray

"Whatever you can do or dream you can, begin it.
Boldness has genius, power, and magic in it!"
Begin it now!

Johann Wolfgang von Goethe

Introduction

I shall weave for you a melody: at times sweet and harmonious, at times melancholic, and perhaps slightly out of tune. Yet always genuine and spontaneous: My African song.

I will allow a tale to unfold that may seem incredible to many yet utterly ordinary to me. A story spun from dreams and adventure, from sacrifices, joy and sorrow, from hope and humility, and above all, from profound love.

What you are about to read is a mere collection of narratives, drawn from the first 25 years of my day-to-day life in Kenya, each of them holding a special place in my heart.

I wished to eternally etch in my memory, and on paper, the vivid images of those long-gone days. They speak of people and places to which I am particularly devoted, and I hope, dear readers, that they will become special to you as well. To truly understand my choice to live in Africa, let me take you back to my early life in Italy. The place where I was born and spent my first 25 years is paramount to understanding the person I have now become. My roots thrived in Italy, and my Italian family (or rather, the remaining members) still holds great importance to me, as it does for my children.

If Africa is my present and, I hope, my future, Italy is my past. Yet, without my Italian past and the people who loved me there, I would not have become the person I am now. I envision my life as being divided into two parts: distinct yet complementary. The first part is my "Italian Life," from when I was born to my mid-twenties, where the protagonist, Antonella, ardently seeks to find herself and her fulfillment; the second is the "African" part, where I discover myself at last and truly start living! This transformation was much like that of a delicate butterfly: from the chrysalis, where the larva grows slowly, preparing to face its future, to its birth, where, unfurling its wings, the butterfly takes flight, embracing life.

While writing my story, I never contemplated crafting the book you are reading now. I solely wished to bequeath to my children a sort of diary, shedding light on how their parents arrived in Africa, and why they chose to bring them into existence on this extraordinary continent. So that, in a not-too-distant tomorrow, perhaps, they too can revel in recounting their own stories. Ones that tell the tale of Africa and Africans, seen through their eyes as "children of the savannah," born and raised among the Maasai in Kenya. So that they too may, one day, sing their African song...

I do not know if this is something my children will comprehend, but I know that writing has helped me in laying bare my essence to understand who I was and who I have become. Africa undoubtedly played a pivotal role in my transformation, and for this reason, I shall be eternally grateful to the continent.

I will not celebrate the Kenya of horrors, tribal wars, corruption, famine, droughts, disease, poverty, and filth in the slums on the outskirts of Nairobi and Mombasa. Much has already been said and written about these terrible realities, although it seems they may never be spoken about enough. Although I know that Africa is also that, this is not the reason I feel inextricably bound to this continent and now consider myself a part of it.

My Africa is one of infinite spaces, one of pure and natural beauty. It is the joy, human warmth, and pride of its people, and the dignity of its animals. It is the Africa of a thousand colors and intoxicating scents. It is the Africa of delicate pink dawns, fiery sunsets, and the vast silver moon.

I sing of its magic, its enchantment, its zest for life, and its extraordinary strength to survive the many adversities that unfortunately torment and threaten the tranquility of its existence every day. I celebrate its wisdom; its ability to teach each of us to live and die, if we are capable of listening.

I celebrate it being "Mother Africa and the Cradle of Humanity," from the beginning of time and forever. I celebrate it being a friend and sister to me, and the "mother" of my children. I appreciate that it never made me feel alone or useless, emptied or tired of living. And never inadequate. I celebrate it always, and only, making me feel a part of the whole, in harmony with life and death, in symbiosis with Nature. In peace with the Universe.

When...

When, at dawn, in Africa, beneath a faint and pale sky, I breathe in deeply the pungent air drenched with dew and the sharp scents of the still slumbering bush. When the first magical vision that unfolds before me is the immense snow-capped Kilimanjaro, brushed, at times, with a delicate pink and a muted blue, its silhouette standing tall against the sky. When I sit on the veranda of my home, and my gaze wanders across the endless expanses before me, dotted with the noble animals of the savannah. When the sky, at sunset, explodes in reds and oranges, and Venus appears and illuminates it, like the fires of the Maasai down on the plains next to the spring. When, at dusk, I hear the breath of the wind gusting from the Chyulu Hills, suddenly breaking on the dry grass of my roof, like a fierce ocean wave crashing on the shore. When, at night, I lose myself in the vastness and mystery of the infinite stars that pierce the African sky, and a huge translucent moon rises majestically from behind the hills, silently floating until dawn in a sea of stars. When, in the dark and silent night, the mighty roar of the lion and the piercing cries of the hyena suddenly echo...

...then, my soul fills with light, and my heart begins to sing.

It sings of this Africa, the one and only that can awaken such emotions in me. It sings of the Africans who, with their patience and wisdom, teach me a little about living and a little about dying each day.

I had never been aware of being "in the right place" until I came to Africa. I had never realized and been in touch with the sense that we all originated from this ancient continent. I do not know how I came to Africa, but I am certain that I will not leave. And being aware of this is what gives meaning to my days.

I am grateful to Africa for opening its generous heart to me and giving me the opportunity to live and love it so intensely. I am grateful to the Maasai for letting me exist amongst them and teaching me to live a new life; more authentic, more genuine, more intense, wiser.

A life worth living.

My Italy

I was born in Brescia, Italy, on the morning of July 11, 1969, at San Camillo Hospital. From the way my mother tells it, it was a hot and humid summer, which was actually out of the ordinary. My parents had longed for a baby, and I arrived almost at once, giving them hardly any time to get used to the idea of having a child. Right after my birth, mom and dad had a card printed on fine ivory-colored stock, in beautiful, elegant handwriting, with the words:

"Sandro and Giovanna joyfully announce the birth of Antonella."

The cards were sent to hundreds of people, friends, family members, and acquaintances, spreading the immense happiness of the new parents.

From what they said to me, I was a beautiful baby – graceful and seemingly serene. I was not particularly rebellious while growing up. Actually, I was always very responsible, with a strong sense of duty, and careful not to hurt or disturb others. I was sometimes melancholy for no apparent reason. I remember being a sensitive child, at times overly so, always carefully considerate. I have never thought of myself as being extremely sweet, or rather, not prone to endearments or displays of affection. Later in life, I would discover an infinite sweetness in my soul, so great and intense that it scared me, to the point of wanting to selfishly hide it in my heart. I would only be able to experience it for a few months at the age of eighteen, after meeting Luca. He was the only one to have found the key to my heart and managed to bring that slumbering sweetness out. It remained dormant in me until the birth of my children, more than ten years later. Much to my surprise and joy, they succeeded in rekindling it when I believed it was no longer possible.

Unfortunately, nothing prepared my family for the emotional earthquake that my choice to live in Africa would cause in their lives

25 years later. There was no sign of my longing to explore another world or my thirst for freedom.

My mother, Giovanna, has always been beautiful. And she was even more so through my young, loving eyes. She has a cheerful, incredibly lively disposition. I have never met a person so enthusiastic about life, so eager to live. I've been told that she laughed often as a young girl. She was fun and very charming. As a child, she told jokes on the beach, in front of everyone. She had no shame and loved to be the center of attention, earning her the title of "family clown!"

She grew up as the middle child between two sisters: the elder one was named Renata, and the younger one Gabriella. My aunts always loved me like a daughter, or even more. I will always be incredibly grateful to them for that.

My mother has always been an exceptional cook. Not only does she have a passion for cooking, but she has always really loved to improvise. With the ingredients she had at home, she could whip up delicacies in short order!

She has also always been known for her great sense of style. When I was growing up, our home was furnished with fine pieces of antique furniture, precious Persian carpets, and paintings by famous Flemish masters on the walls, carefully chosen by both my parents, avid antique aficionados. Beyond this, our home was also welcoming and comfortable, always tidy and cozy. It was decorated in a truly special way that was like an invitation to stay. Without a doubt, this was thanks to my mother, who was praised for her exceptional taste and hospitality.

I always remember her being very elegant. Tall and statuesque, anything she threw on looked graceful. Her straight blonde hair framed a face illuminated by two deep blue eyes, though they were often dimmed by a shadow that I could not quite understand. As a child, I wondered what dampened my mother's spirit, what sometimes stole her smile and her joy for life.

Unfortunately, despite her cheerful and outgoing nature, I remember my mother as a woman who was often sad and fearful, especially when at home with my father. Though I know she was also very much in love with him. An authoritative man, he was deeply unhappy and dissatisfied with his life, but not strong and mature enough to do something about it. My father was so unpredictable, so excessive... when it came to his feelings and throughout his life. My father was desperately depressed at times and madly euphoric at others. In search of himself and his happiness, my father was a man who never was able to love himself and thus could never truly love others. Except for his mother. And perhaps, in his own way, his Antonella, for whom he had grand plans, hoping that she, one day, would be able to accomplish what he did not.

My mother has told me that my father adored me when I was little. He played with me for hours; always very sweetly and affectionately. Then as I began to get older, something changed between us. Or perhaps, more simply, he changed. During my childhood and, more frequently, during the challenging teen years, I clashed with his pessimism and anxieties. It wasn't easy to live with him while also maintaining a sense of equilibrium and trusting in the future. I am convinced that my decision to go to Africa also had to do with "escaping" from him, and my desire to start a new life away from his dark, sad world.

When I decided to go to Africa, my mother and father separated. I didn't think it was a coincidence as I had been the glue that had held them together. When I left, everything collapsed and crumbled. But it was for the best. After they parted ways, my father seemed more serene and perhaps, if possible, a little happier. My mother was reborn; or rather, she came back to life. That shadow of sadness I saw in her beautiful eyes when I was a child went away. She is an attentive and loving mother to me and a wonderful grandmother to my children.

My sister Chiara was born when I was almost four years old. We looked a lot alike physically as children, but her personality was the opposite of mine. She was impetuous, lively, capricious, extroverted, and very charming. Because she was so exuberant, she immediately attracted everyone's attention in the family. Like most of first-borns, I was very jealous of her at first. Then I saw her as my little sister, in need of protection and love. Chiara was very attached to me and often sought refuge in me when she was sad or had problems at school or with friends. For her, I was not just a sister, but someone to count on, a role model to follow, and a guide in her young life. We would play together for hours and always kept each other company. I don't remember ever having a serious fight with her. Throughout our lives, we became not only sisters but also good friends, always supporting each other regardless of our differences, which may actually be the reason we are so fond of each other.

I grew up surrounded by love and a little bit spoiled, especially by my aunts and grandparents. Unlike other members of my family, as a child, I was introverted and kept my emotions to myself. I was almost ashamed to show my feelings and to open my heart to others. I often was cloaked in a sort of "armor" made up of fake cheerfulness and indifference, to not show others who I really was, to not get hurt, and thus, to not suffer. Writing has always been the way I show what I truly have inside my heart. Only with a pen in hand, before a sheet of paper did I become truly honest; first with myself, and then with others.

In my family, the person I was most like in this sense was my aunt, "Zia Renata," my mother's older sister, with whom I have always had a particularly deep bond. Zia Renata was the intellectual of the family. She read a lot and loved to write. Often, when we were alone, she would read her diaries to me. I found them moving and they always warmed my heart. She was, and still is, a true friend and my go-to person. Even now that she is gone, she remains alive in my heart and always present in each moment of my life. I miss her immensely, but I know that she is always with me.

Renata was the beauty of the house. She was also blonde, with piercing blue eyes and a beautiful smile. I always remember her as cheerful and radiant, very sophisticated, and a woman of great class. She had great strength of character and profound wisdom, even from a very young age. At the age of 20, she fell ill with a kidney problem and was on death's door. Her luck and strength got her through it, but not without a long, painful post-surgery recovery. After the surgery, Renata had to stay in bed for several months as she recovered. I believe this matured her and taught her the true value of life. Actually, it may have been these health problems that led to her being so wise and aware.

Zia Renata's pragmatism made her a key figure in the family for solving all of those problems others found too daunting. As children, when we were absolutely desperate and could not find solutions to our seemingly insurmountable problems, we would run to Zia Renata for her valuable advice, certain that she could help us! And that's how it always was.

Zia Gabriella was the "little" one in the family and, therefore, the most pampered by everyone. Very delicate in health, of the three sisters, she was the one who physically resembled her mother the most. Despite her sweet, fragile appearance, she was the strongest of the three sisters and the most fearless. She was undoubtedly the most stubborn: if she got an idea in her head, it was impossible to change her mind! She was also the most adventurous. At 20, she left her home and the comforts it represented to live in Brazil with her boyfriend.

At that time, in a society that was bourgeois and somewhat judgmental, that decision caused quite a stir. As always, Zia Gabriella didn't care about what people thought and followed her love, Federico. A geologist from Brescia, he was unique and certainly a very charming person. He was obsessed with searching for gold but had never found it! He lived in Nigeria for a while then decided to try his luck in Brazil. Gabriella, madly in love with him, followed him and embarked upon an unforgettable adventure with him.

They exchanged vows in São Paulo, under the watchful eyes of her father (the only member of the family with the courage to embark upon the long journey to Brazil!) and two witnesses they barely knew. To marry the man she loved, Zia Gabriella had no need for a ceremony with relatives and friends or even a grand celebration. She detested formalities and appearances; her wedding, which was hastily celebrated in a foreign land with two unknown witnesses serves as the perfect testament to this.

Zia Gabriella had always been my confidante. When I told her a secret, I could be sure that she wouldn't betray my trust. I loved being in her company. She was incredibly open-minded, perhaps due to her years abroad, and we conversed about a myriad of things, always deep and unexpectedly thought-provoking. Because we were both strong-willed and firm in our convictions, we often had quite spirited discussions. I cherished those moments of complete trust and exchange. I have always been convinced that they helped me mature and grow as a person.

I had two cousins – the children of my mother's sisters – with whom I was particularly close in childhood: Giacomo and Nicola. Giacomo, who is four years older than me, was the son of Renata and Marino. He was a handsome boy with dark eyes and a bright smile, lively and always ready to teach me adventurous games. When we were little, we wanted to get married, until – to our great disappointment – our parents explained that we couldn't because we were related.

Growing up, Giacomo became a sort of older brother, always protecting me and taking care of me. Giacomo could truly be my brother: he is the son of my mother's older sister, Renata, and my father's older brother, Marino. Yes, these two brothers actually married two sisters! My cousin is a successful man, deeply committed to his work and his family. He is always at the ready to solve everyone's problems and is often called upon for advice or to mediate delicate situations. He often visits me in my Africa, which is now near and dear to him as well. I care for him deeply, and the awareness that I can rely on him for the rest of my life warms my heart.

Nicola, the son of Gabriella and Federico, is six years younger than me. He has always been closer to my sister Chiara, probably due to the significant age difference between us. He was a beautiful child – lovable and very intelligent. He was very sensitive, almost too much so at times. He played with Chiara for hours on end without ever fighting with her, docile and sweet, smiling and cheerful. He had a keen ear and a great talent for music. He could play almost every musical instrument without ever having been taught. His favorite was the piano, which he played for hours, singing well-known songs as well as others he had composed himself. I know he would have become a talented musician if he had been able to find his path.

During his teenage years, he changed and drifted away from us. Except for Chiara, our group of cousins no longer understood him. Unfortunately, we slowly grew further apart from him, leaving him in the darkness of his solitude. The only person who was and is still always by his side – with determination and immense motherly love – is Gabriella. Even now, she is the only one who holds the keys to her son's heart and can glimpse, at times, the faint colors of his troubled soul. We hope that one day, he may return to his cheerful self and the joy we once shared can be found again.

I always preferred being alone to being with others, especially as a child. I was very independent from a young age. My mother said she would sometimes forget that I was at home, as I was so well-behaved and quiet. My little room was my entire universe, and I was perfectly happy there. In that room, I felt safe, protected from the rest of the world. Perhaps for that reason, I didn't like going to preschool, where I felt obliged to be with other kids and to "socialize." Fortunately, I barely ever went! When I turned six, and it was time to go to primary school, I went willingly and never had any problems integrating, making many friends along the way. Over the years, I became a good student, held in high regard by my teachers and a source of great pride for my parents.

I grew up experiencing things typical for girls my age; believing I was happy and diligently doing what others expected of me. Over time, I would let go of the shyness and reserve that had characterized my early years. As a teenager, I became extroverted, assertive, and exuberant. A girl with great strength of character, determined to get what she wanted.

My life as a "good family girl" unfolded seemingly peacefully and predictably, much like the lives of most people my age and in my area. I chose to study foreign languages in high school and then enrolled in law school in Parma, where I lived my university years with a dear friend in a small apartment in the center of town. I had some of the most carefree, wonderful moments of my youth in Parma. After graduating, I chose to run the family winery instead of practicing law. I liked it, and it felt like I was continuing what my ancestors had started. That job gave meaning to my life, and that was enough at the time.

Despite leading a relatively trouble-free and privileged life, I sometimes felt like I wasn't truly satisfied. I felt that something was missing, without truly understanding what. I felt guilty for these thoughts. I realized how lucky I was and how many other people I knew were less fortunate. Nevertheless, inside of me, I felt a deep sense of inadequacy, almost impatience, toward the life I was living and was expected to continue. Inside of me, I now understand that there was a lingering restless spirit, yearning for yet another life, another world. Now I know that spirit would only find peace in Africa, but that was a few years off.

Italian Holidays

PONTE DI LEGNO

My most cherished childhood memories, painted with an Italian brush, are inseparably woven with my grandparents and their holiday homes. Those were the times when we, as carefree children, spent our summers without fully picking up on the extent of our good fortune.

I have always had a great attachment, almost a physical bond, to our family homes and their contents. From the majestic furniture to the expressive paintings, the intricate carpets to the tiniest of treasures – a whimsical trinket, a frozen moment in a photograph, a book, or the sets of dishes and glasses – these homes, and consequently, the objects in them, tell the stories of the people I hold dear. I have a fondness for old things; those objects have been part of my family for as long as I can remember. I relish touching them, breathing in their essence, and listening to their silent tales. In doing so, I see the echoes of those who came before me. It is as if I can still hear their distant voices, catch the faint whiff of their familiar scents, and feel their gentle caresses.

When I see my children in those houses, stretched out on the same sofa where my grandparents used to sit or dining at the same table with the dishes they once used, a pang of bittersweet nostalgia washes over me. Yet, there is also immense joy and gratitude for the ability to continue living while honoring those who paved the initial path of my life. These "things" are the delicate threads that bind me to my departed loved ones – silent shadows that visit me during my lonely evenings yet remain profoundly etched in my heart.

In essence, I believe this is the true meaning of life – we are "passed down" from one generation to the next even through inanimate objects. It is about not losing ourselves and not forgetting those who came before us. It is a path without a clear beginning or end, much like an infinite loop or a circle of life where energy renews itself through the tapestry of our lives. Forever.

When we were young, our parents would sometimes take us to Tuscany, often in July, to a small, welcoming hotel in Forte dei Marmi mere steps away from the sea. This lively seaside haven was the perfect playground for our childhood adventures. I have fond memories of the long bike rides through the pine forest with my friends, indulging in cream-filled pastries on the beach after an afternoon swim, boat trips with my father teaching me to sail, and the lively Wednesday morning market. I remember holding my mother's hand in the crowds of shoppers that swarmed the stalls brimming with the exquisite Florentine linens and other treasures that piqued my childhood curiosity.

Yet my favorite holidays, which are still vivid in me, were the ones in the mountains in August or at Christmas, in the Italian Alps in Ponte di Legno, at the vacation home of my maternal grandparents, Giacomo and Vittoria; or in the countryside, in September, at Castellino, the home of my paternal grandparents, Sebastiano and Nella.

The Ponte di Legno house was beautiful and harmonious, built like most mountain homes in wood and stone. Cozy and comfortable, it was nestled in dense forests of pine, fir, and larch. The dining room windows, adorned with pots of red geraniums, looked out at the sharp peaks of Castellaccio and the Adamello glacier, which were covered in snow even in summer. My grandfather Giacomo, a fervent mountain lover, had the house built in the 1950s when the surroundings were just meadows dotted with colorful mountain flowers and expansive green valleys. There, in the summer, cows and goats grazed peacefully, coming down from alpine huts.

The house was quite large; at least, it seemed enormous to me, and it was big enough to accommodate – in addition to my grandparents – our family, Zia Renata's family, and Zia Gabriella's family. I remember the cool summers and the white Christmases in Ponte as the most beautiful vacations of my life, filled with love, warmth, joy, and carefree moments shared with the entire family. During the summer afternoons, we children played hide and seek in the woods below the house. I still remember the scent of moss, the coolness, and peace of that forest, and the crunching of pine needles and cones under our feet. Even today, when I come across a similar scent, my heart smiles, and it takes me back to the innocence of those early days.

During the summer, we took long walks to see the agile chamois, a type of goat-antelope, leaping from one cliff to another and the majestic eagles soaring high in the sky above. We would often take a break to enjoy a picnic by a babbling brook. After lunch, it was siesta time for the adults, while the children lay in the grass in silence, beneath the shade of towering pines populated by squirrels and graceful tit birds. We gazed at the sunlight filtering through the branches and clouds, running light above us, soft and white like cotton candy. Those moments of silence were filled with unspoken words and pure, genuine family love.

In the winter, our days were spent on skis. We would excitedly wake up in the morning, overcoming any lingering laziness and the biting cold on our hands and feet. Still half-asleep, we would clumsily slip into our ski suits. After breakfast, we would head to Tonale and ski until the late afternoon. The exhilarating sensation of descending swiftly down snow-covered slopes could not be beat! Sometimes, we were fortunate enough to ski while it was snowing – everything turned white and silent, pure and calm. Returning home after skiing, we would gather in the living room, huddle in front of the fire and sip hot chocolate. Grandmother would read us a story or simply listen, her eyes full of love and interest, to the tales of our skiing adventures.

In the evening, we would all gather around the old walnut table for dinner. We were always a large group. The women would cook, helping and chatting merrily among themselves. The men played cards, and grandfather would read the newspaper in his armchair. If she wasn't in the kitchen, grandmother would be embroidering. The children set the table before dinner and cleared it after we finished eating. It was a kind of ritual that never changed and made us feel part of something that gave meaning to our young lives. I remember those dinners, all of us together, as rare and precious moments of joy to be cherished in my heart forever.

Grandfather Giacomo and Grandmother Vittoria were an unusual couple. He was a handsome, elegant man who captivated the attention of many women. Tall, blond, with intense blue eyes, much like those of his daughters. A successful businessman, his employees greatly admired him and held him in high regard. Although he lost patience easily and used rather colorful language at work, he was incredibly generous. If his employees needed help, even financially, they would approach my grandfather, knowing they would not be turned away. Ever.

Giacomo adored opera and introduced it to his grandchildren from a very young age. Every season, he got a box at Teatro Grande, never missing a premiere or an evening opera that he particularly liked. In the early years of their marriage, Vittoria accompanied him, looking ever so gracious in her evening gowns. However, she struggled to be in the midst of so many people, smiling and conversing. She was always very shy and reserved. Over time, Giacomo started taking his daughters to the opera house instead of his beautiful wife, especially my mother, who inherited from him a love for opera and socializing. She was proud to enter the theater on her father's arm, whom she admired and loved deeply.

Grandfather was always impeccable in his elegant city attire, but in the mountains, he wore soft velvet pants, plaid shirts, and comfortable cashmere cardigans. This is where he could finally relax! In the summer, he rose very early. He loved to hunt for mushrooms in the woods and was incredibly skilled at finding them. We grandchildren fought to go with him. His gray poodle, Ocio, was always by his side and faithfully followed him everywhere. They were an inseparable pair, and when Grandfather passed away, Ocio stopped eating and faded away in a matter of days.

Grandfather could be authoritative and strict with us grandchildren at times. We respected him greatly and were somewhat afraid of him. Yet, he often joked and could be very kind, playing with us and making us laugh. He never wanted us to call him "grandfather." He only wanted to be called "Uncle Giacomo." He was certainly an unusual grandfather! Perhaps this is why he was so dear to us.

Vittoria, on the other hand, was incredibly sweet and gentle, the most accommodating person I have ever known. She had brown hair, with large, kind dark eyes and a beautiful smile. Reserved, silent, and shy, she loved poetry, reading, and classical music. I never heard her raise her voice with the grandchildren or her daughters, whom she loved more than herself. Grandfather was the only one who could truly get a rise out of her! My grandparents loved each other deeply, yet their essences were so different that coexistence became a delicate dance. With her sensitive spirit, I believe my grandmother often felt hurt by the impetuousness of my grandfather. They lived separately for a period, but after a few years, grandfather retraced his steps home to his sweet Vittoria. Together, they weathered the tapestry of life down to its final thread, filled with love, hard-earned respect, and each other's company.

In her youth, my grandmother graduated from music school with top honors. She played the piano splendidly. It was almost an extension of her being, with her face changing expression as soon as her fingers touched the keys. She detached herself from everything and everyone – it was as if she ascended to a surreal dimension, filled with magic and pure harmony. Unfortunately, she was so shy that she never played in front of anyone. She didn't play for others; she played for herself.

In the afternoons, she often retreated to grandfather's study, which smelled of leather and tobacco, where the imposing grand piano was located. Here she would begin to play. Chopin was her favorite composer, with his poignant sweetness and melancholy. As a curious child who adored music, I would prop myself sneakily outside the door in reverent silence, enchanted by that melody.

I never told grandma, not even when I got older, that I lingered outside the door and listened to her gracious melodies. Who knows if she knew and pretended not to? I should have told her. I am sure it would have pleased her. Wherever she is, I am sure she knows now and smiles at the delicate thought of it.

When I turned 10, grandma decided to try to pass on her gift for the piano. Though I had always loved music, the piano was not my forte. I found myself drawn to the classical guitar instead, and I quickly learned to play, singing along to the music.

Aside from choosing a different musical instrument, I must say I am much like my grandmother. Today, those who know me say I play the guitar and sing very well. However, like grandma, I do it mostly alone; almost never in front of anyone. Playing is an intimate affair for me, shared only with those to whom I am particularly close.

I played for Luca at the beginning of our relationship, and he fell deeply in love with me as he listened. Now, I rarely do it, despite my husband's continuous requests. If I play, I do it for my children. In my daughter's deep, intelligent eyes and in her enchanted, captivated expression, I see myself as a child, hidden in a corner, listening to grandma – who was perhaps unaware of my presence – play for hours, admiring her skill all the while.

Grandma Vittoria loved the snow more than anything. When I was 16, she fell ill with intestinal cancer and died quickly, after a great deal of suffering. She departed this earth one morning in early January. A few minutes after she passed away, pure and light snowflakes began to fall gently from the gray sky, as if to cover all the pain and suffering she had endured. As we gathered around her bed, we looked at each other in silence, feeling moved. Grandma would have been happy to see it snowing. Her suffering ended, and she was finally at peace.

All of us felt great sorrow at her passing. Chiara, in particular, suffered a lot. She had always been very close to grandma. One could say she had experienced the most beautiful days of her childhood with her.

After grandma died, I often went into her bedroom. I missed her very much. It was as if I could still smell her sweet scent. There, surrounded by her things, my wounded heart calmed. One day, as I looked through her dresser, I moved some silk scarves in a corner and found her diary, written in her clear and elegant handwriting. I sat in her favorite armchair, next to the window overlooking her beloved garden, and feeling deeply moved, I began to read:

"Do not let a day pass without discovering in it a secret, be it large or small."

And again:

"It is not you, sweet, beautiful red branches swaying in the wind: it is my soul."

These were her verses, but there were also passages from poems she had particularly loved and diligently transcribed. It was then that I realized my grandmother was not just the sweet, serene woman I had always imagined, but a spirited soul with a fierce heart, overflowing with passion and vitality. From that moment on, I loved her even more.

Six months after my grandma's death, grandpa Giacomo fell ill. I believe he allowed himself to die. Without his Vittoria, living no longer made any sense.

The death of my grandparents was the first great, true pain of my life. One of those immense absolute aches that enter your soul and stay there forever. One of those heartaches you never forget. Something died forever in the hearts of my mother and aunts after their parents died. But isn't this the circle of life? The first scars of suffering, caused by physically losing a loved one, teach you to live and, perhaps, a little bit to die. This is what Africa taught me many years later.

Today, Grandma Vittoria's diary is always with me, on the bedside table near my bed in Africa. When I miss her, I pick it up, caress it, and reread it. This way, I feel her close to me again, and my heart seems a little lighter.

The Castellino

The Castellino was purchased by my paternal great-grandfather, Fausto Bonomi, in the early 1900s. Fausto was a businessman who often traveled by train from Brescia to Milan. The train passed near Coccaglio, under Mount Orfano, a solitary and proud mountain that rises in the southern part of Franciacorta, about 20 kilometers from Brescia, in the northern part of Italy.

From the train window, right at the foot of the mountain, Fausto always looked at the Castellino, a truly unique building that he grew to love dearly. It was a small castle in Liberty style, like the Austrian castles of Ludwig, built at the end of the 19th century by the famous architect Tagliaferri. Surrounded by elegant boxwood hedges, it was immersed in a garden of ancient trees, inhabited by deer, peacocks, pheasants, and hares. It overlooked sunny vineyards, neatly arranged on the steep slopes of Mount Orfano. The ancient castle entered Fausto's heart with force, and he wanted to buy it as a gift for his lovely bride.

One winter morning, during one of his many trips back and forth to Milan, while calmly contemplating his upcoming appointment on the train, Fausto comfortably sat back on the red velvet seat. Across from him, a big man with a striped waistcoat and a watch with a long flashy chain blissfully smoked a cigar, casually gazing out the window. After stopping in Coccaglio, the train headed towards Milan leaving Mount Orfano behind. The mist covered the vineyards around the little castle, blanketing it in a mysterious haze. Fausto silently admired the Castellino with dreamy eyes as it appeared suspended on the mountain, basking in the sun. The big man across from him looked up and said, "It's for sale, you know!" My great-grandfather hurriedly got off at the next train stop, Palazzolo, forgetting everything about his important appointment. He flagged down a carriage and sped towards Coccaglio, while his heart sang, "I'll buy it!"

Since that time, the Castellino has always belonged to our family. Fausto's brothers, children, and grandchildren often visited, especially during the long, warm Lombardy summers. When great-grandfather Fausto passed away, he left it to his son Sebastiano, my paternal grandfather, who lived there, loving it just as his father did. He stayed until his dying breath, alongside his beautiful wife Nella and their three children: Marino, Sandro (my father), and Fausto.

The three young, vibrant boys lived with their parents in Brescia and stayed at the Castellino only in the summer. However, during the winter weekends, they would secretly hold wonderful parties at the Castellino; the house was large and perfect for their fun fetes! In no time, the Castellino became well-known throughout Brescia and the province thanks to the splendid parties held there. Sebastiano, realizing this and seeing his sons neglecting their studies and having too much fun, decided to sell the property. He impulsively sold it, driven by anger and feeling deceived by his sons. But he regretted it immediately. Fortunately, he managed to buy it back within a few months, just as his father Fausto had done many years before. Perhaps it was destiny for that house to remain in the Bonomi family...

The Castellino was my first home. After I was born, my parents, who lived in Brescia, decided to move to the countryside with me. Because it was July, it was very hot in the city. In Coccaglio on Mount Orfano, where the Castellino stood, the climate was cool and breezy. Every day, my mum took me for long walks in my stroller through the vineyards, with their neatly arranged, well-tended rows of vines. Sometimes, she would lull me to sleep in the shade of the enormous cedar trees in the park behind the house. Grandmother Nella and Grandfather Sebastiano also stayed at the Castellino with us; they cuddled and spoiled me every day. In October, we all returned to live in the city. The Castellino was a vacation home. It would remain closed until the arrival of the next beautiful season, spring. A lovely attic in the historic center of Brescia would be my home until my departure for Africa, 25 years later.

From a very young age, Chiara, Giacomo, and I went to the Castellino every year at the beginning of September. We stayed with our grandparents until the beginning of October, right before the start of school. With the warm, golden colors of autumn painting the vineyards and the mountain like a French impressionist, it was the most beautiful time to be there. We often played in the farmhouse for hours during the lazy afternoons while our grandparents rested and the cicadas chirped in the fields.

We would climb up a wobbly ladder to the attic of the hayloft. From there, alongside our little friends—the farmers' children—we would leap onto the hay bales, our laughter ringing through the air as we tumbled down. With each soft landing, the scents of the sun and the countryside rose around us. I still remember the dust in my nose, our red eyes, and the hard, sharp hay poking through our light clothes, stinging the skin of our legs and arms. We silently explored the stables and the henhouse, befriending all the inhabitants. Our favorite game was releasing the newborn rabbits from their cages when Angelo, the farmer, turned his back; they were so sweet and fluffy! I nostalgically remember breathless races in the courtyard and bike rides along the stream nearby. We were happy and free in a way we haven't been since. Our tennis and bocce games often resulted in spirited shouting about someone cheating and, therefore, winning the game unfairly!

And then there were walks up Mount Orfano, to Luigi's (Angelo's son) hunting lodge, the rookery, as he called it. The lodge was a secret and enchanted place, populated by a myriad of little birds. There were so many times that I – unsuccessfully – tried to convince Luigi to stop hunting. But those were different times when hunters were often so enamored with nature that they found that hunting was seemingly the only way to possess it, like a beautiful lover that slips away when you want to make her yours.

Often, in the afternoon, Grandma Nella took me to pick fresh flowers in the meadows next to the house. She would then take me by the hand and lead me to the little church behind the Castellino. Once there, we would place the flowers in a crystal vase in front of the Madonna statue and would light a candle. We would recite a prayer for our loved ones in the cool, serene silence. We would blow a kiss and return home with a sense of calm and peace.

The most wonderful moment in the countryside, which we children always eagerly awaited, was undoubtedly the grape harvest in early September. We would go into the vineyards to pick ripe grape clusters at dawn. The farmers would give us wooden stools and rusty pruning shears to reach even the highest of grape clusters. We made it a competition to see who had harvested the most by the end of the day. Once the harvest was over, the most fun part began. Angelo made us stomp on the grapes, barefoot in huge wooden vats, until we collapsed exhausted in the evening, falling asleep under the farmhouse porch, lulled by the women singing and the crickets chirping.

Grandpa Sebastiano started producing red wine more as a hobby than an actual profession. He used the Merlot, Cabernet, Nebbiolo, and Barbera grapes that already grew on the property, and the wine was meant for family consumption only. Uncle Marino, Sebastiano's eldest son, is the one who later turned this hobby into a real agricultural business and, therefore, a successful endeavor.

It can be said that Zio Marino was the true "heir" of our grandfather. At least, he truly inherited the gift for the art of winemaking, something that had been a source of great pride and happiness for his father. Marino, an engineer, was intelligent and honest. In addition to his career, he devoted his heart and soul to wine production, which he discovered to be his true passion! Zio Marino was kind, gentle, and always willing to help us children when we needed it. He had a great love for sports, racing cars, the countryside, and his wife Renata, an immensely important figure and rock for him. I was very fond of Zio Marino: he was, and still is, a very important person to me. He is the person I turn to when I need advice about complex decisions in my life. He is a wise and optimistic man, and a lover of life; his soul is filled with light and beautiful energy.

In addition to the existing vines on the property, Marino had Pinot and Chardonnay grapes planted. Surrounded by expert consultants and agronomists, in no time at all, he expanded the production of Tenuta Castellino Bonomi. In addition to the original red wines, he began making fine still white wines, and, most of all, excellent sparkling wines! From a young age, Zio Marino took us to visit the cellars, and I still perfectly remember the scent of wine and must that seep into the damp walls. He taught us, with seriousness and passion, the first rudimentary art of wine tasting: *"First the color, then the aroma, and finally, the taste,"* he would say seriously. He hoped that at least one of the grandchildren would become passionate about the art of winemaking.

I believe that if I hadn't decided to go to Africa, I would have stayed to work at the Castellino, with the same passion and dedication that Zio Marino has.

Unfortunately, I have few memories of my grandfather Sebastiano. Even today, he is remembered for his sharp business acumen, his honesty, his decency, and his infinite sense of duty. He died when I was only 13, so I do not think I had time to really get to know him. As a young man, Sebastiano was shy and reserved, very polite and serious. But he was also a man of strong passions. At only 20 years old, he fell madly in love with Grandma Nella, his "Nellina," as he called her, on the very day he met her in Lumezzane, their hometown. When they met, Nella was very young, just 16 years old. She was beautiful, unpredictable, charismatic, and fascinating. Sebastiano was only a few years older. He made her the center of his life and the queen of his shy heart until his death. They married at once and soon had three children.

My grandmother Nella was special to me. I was named Antonella in her honor; fatefully, our common name strengthened the deep bond that would develop between us. She was friendly, original, outgoing, and fun, and much younger than her age. She loved jewelry, and she had beautiful pieces that she would bring out for social occasions but also for everyday. She was very generous with everyone. There were times I told her I liked a particular piece, and she would take it off and give it to me immediately even if it was precious. She would say, *"Take it. It's yours! And always wear it because jewelry is made to be worn, not to be abandoned somewhere. Jewels are alive and need love and light, like plants or flowers..."*

Grandmother had a gift, an ethereal knowledge. She had a "sixth sense" for how events would turn out. She claimed that she could communicate with her mother Maddalena, whom she dearly loved and who succumbed to cancer when Nella was just 13, through a golden pendulum. This precious, perfectly round jewel adorned with intricate floral and leaf motifs hung delicately from a slender chain around grandma's neck. On certain "special" days, she would steal off to grandpa's study, remove the chain from around her neck, dim the lights, light a candle, close her eyes, and launch the pendulum's rhythmic sway, back and forth... depending on the direction it began to turn, grandma provided answers to our questions. Whispers like, "If Antonella's math test is going to go well tomorrow, swing to the right... "If grandpa is going to recover soon, swing to the left..."

I always suspected that she subtly and imperceptibly moved the pendulum herself, so she was the one who gave me the desired answers. When I expressed my doubts, my grandmother would get extremely angry. She insisted her hand was perfectly still, and the pendulum, moved by the energy of her departed loved ones, responded on its own. Over the years, I often found myself questioning her about the future outcomes of school tests and university exams. Sometimes, I'd seek advice about my unhappy teenage romances as well! I always left feeling calmer after consulting her, and generally, what she predicted came true.

Unfortunately, Grandma Nella too fell ill with cancer and passed away when I was 18. In the last months of her life, when she was confined to bed in pain, I visited her every afternoon after school. I brought my guitar, which she adored, and played for her. I saw immense love for me in her tired, suffering eyes as well as a hint of anger about life, or rather, the death that was coming to take her away. I believe she deeply regretted knowing that she wouldn't see me grow up and wouldn't meet my children. Nevertheless, towards the end, she told me one day that she was certain I would lead a happy, fulfilling, and extraordinary life and that I would have three children. She said the pendulum had told her! It couldn't be wrong...

None of my grandparents ever knew that I decided to move to Africa because they all passed away before I left. I wonder what they would have said, how they would have reacted to my decision. I know that no matter what they would have understood me and continued to support and love me. Perhaps, even more than before.

Reflecting on my years in Italy, with the wisdom of hindsight, I now know that meeting Luca in Brescia when I was 18 was the catalyst for change in my life. Luca, who became my husband several years later, inspired me to seek out a new Antonella, to embark upon a new path, and to live a new life. He gave me the courage and strength to try.

At first, he swept in like a hurricane, bringing upheaval to my life and the lives of my entire family. Wanting to take me to Africa meant taking me away from them. Forever. It wasn't easy for my relatives to learn to love him and accept my decision to leave Italy. It wasn't easy for those who loved me so much to get used to living without me.

When, in 1996, after years of emotional turmoil, I decided to leave permanently for Africa with Luca, Zia Renata, overwhelmed with the pain of my departure but also filled with all of her love for me, wrote to me, *"Only those who dare can fly."* Even more than I did, perhaps she understood that I was ready to spread my wings.

Luca: Love and Freedom

I met Luca, who is now my husband, outside my high school in Brescia. It was June 1988, and I was preparing for my final exams. A friend who knew him well kept telling me that he would be the man of my life, that I absolutely had to meet him, and that we would get married. She said the same to him about me. But neither of us believed her, not even for a moment.

You could say that it was "love at first sight." It felt like there were sparks between us and we had an immediate strong physical connection. We were drawn to each other in a way that struck us deeply, changing our lives forever.

Luca was bold and charming, yet sweet and dreamy. His soul was filled with passion, light, and energy. He loved his freedom and Africa equally. In my eyes, he was courageous and adventurous, with a strange ability that would become one of his greatest qualities over the years: the capacity to be a leader. He had a remarkable ability to persuade others of his beliefs, and his passion for his dreams was infectious, even to the most rational minds, who were typically resistant to such fantasies. He was the first man who made me feel like a woman, and for that, I fell madly and completely in love with him. In his strong arms, I found refuge and felt protected. I felt like I was at home.

When I met him, he was tan, and his vivid, deep blue eyes stood out on his handsome face. They reminded me of the turquoise sea of Sardinia, which I loved more than anything. He was undeniably seductive; intelligent, charismatic, and witty. I remember the impression I had of his hands. His fingers radiated sensuality, energy, and determination. He carried with him the scent of the golden grass of the savannah, the desire for adventure, boundless horizons, and the infinite starry skies of Africa. Being with him made me want to dream, to dare to fly higher and higher.

I was 18, and Luca was five years older than me. That seemed like an eternity to me. He was studying economics and finance, though he was not particularly dedicated or committed. His father had insisted that he enroll at the University of Brescia after high school, forcing him to continue his studies before carrying out his plans to move to Kenya. Having promised his parents that he would graduate, only after obtaining his doctorate could he decide what to do with his life. But his heart was already in Africa, it had been for a long time...

The day after we met, he came to pick me up after school in the car of my dreams: a white Alfa Romeo Duetto convertible. He pleasantly surprised me with a picnic basket containing all sorts of delights! I couldn't have known, right there and then, how many more times throughout my life he would surprise me. And with the words, "*Let's go, I'll take you to lunch in a special place!*" our adventure began.

It was a beautiful June day. The crisp, cool wind played with my long blonde hair, catching the sun shining above us. In the convertible, we sang love songs by Mina, our favorite singer, so happy and carefree. Luca skillfully drove on a winding road until we reached the top of Maddalena, the hill overlooking Brescia.

We settled on a blanket on the grass scattered with white daisies, in a place with the absolute best view of our city. It was the first time we were alone together. I was struck by how comfortable I felt and how much we had to say to each other. It seemed like we had known each other forever. Now I know our souls had already met and loved in another life, and we would be soulmates forever. That day, as Luca was lying with his head on my lap, looking at me seriously, I got lost in his eyes. He talked about his love for Africa, and I listened, enchanted and captivated by the intensity of his feelings for that distant and fascinating land...

He had dreamed of Africa since he was a child. To get in touch with the feeling of being in the bush, at only four years of age, he slept in a little tent in his garden in the Lombardy countryside. He stayed out there in complete darkness and solitude, with no one except for his trusted companion, his dog Cochi, showing what a brave child he was!

Even then, he began to prefer the company of animals to being with people.

His father, suffering from what we call "mal d'Africa," (a profound yearning for Africa) had left his heart on that special continent, and had taken his son on many African safaris. Luca later told me that he had fallen more and more in love with that ancient and mysterious land, its people, and the wild animals that inhabited it, so much so that he kept going back every year, looking for an opportunity to live there permanently.

That day in June was the beginning of everything. And from that day on, my life changed forever.

We dated for a few months, riding the wave of intense emotions, deep conversations, magic, poetry, romantic nights under the stars, and African dreams. However, despite the strong feelings that bound us, Luca went his way, and I went mine.

We continued to love each other silently, and every time we met, it was as if we had never parted ways, as if nothing had affected what we felt in our hearts for each other. We would continue to reconnect and then drift apart until we decided to be together permanently six years later. I believe that when we first met, we were not yet ready for something serious and final. I am sure we subconsciously feared tying ourselves to each other, definitively, and inextricably. I have always considered this to be a great asset to our relationship.

The fact that we sought each other out for so many years and finally found each other was a sign that destiny had guided us and brought us together, that our love was written in the stars. It was beautiful to have waited so long before being able to truly say that we belonged to each other. Today, we appreciate what we have, and we are aware of the profound, mature bond between us.

In the meantime, my life in Italy continued. After graduating with a law degree, I started working in the family vineyard in Franciacorta, surrounded by barrels and barriques, with a sense of commitment and satisfaction. I had fleeting love affairs that lasted only a moment, not realizing that, in my heart, I compared every man I met to Luca. They could never quite live up to him!

During our years apart, I would meet him again, my love, on my path, from time to time. We would find our way back to each other, and every time we felt closer and closer. They were short moments, filled with emotional intensity that shook us and scared us, making us run away from our love. Continuously.

Until something happened that was terrible but extraordinary at the same time, that made us understand how important we were to each other and made us finally decide to be together. Forever.

The Accident: a New Antonella

It was July 11, 1994, my 25th birthday.

My sister Chiara was driving my car, a Volkswagen Passat, towards Forte dei Marmi in Tuscany. Chiara, young and inexperienced behind the wheel, had recently gotten her driver's license.

It was a beautiful day; the sky was blue, and the intense sun was high above.

The night before, I had thrown a party in the countryside to celebrate my birthday with friends, music, and too much wine! After the last guests had left, I remember spending the rest of the night with Luca, more intensely and passionately than usual, as if subconsciously aware that it could be our last night together. At dawn, as we said goodbye, I told him, *"You know, we should experience what we have deeply. Our love is unique. We have a duty to protect it. We are lucky. If I were to get in an accident tomorrow and not be here anymore, I wouldn't forgive myself for not having fully experienced our love. Maybe it's time to stop running away from it."* I don't know why I said something like that. It was strange for me to express these thoughts; I had never said anything like that before, and I was shocked when the words came out of my mouth.

I shivered thoughtfully in the morning breeze as I got into the car to go home and finally head to the sea.

That morning, after all the revelry, I was still sleepy and asked Chiara to drive. I usually didn't ask her to drive, but just that once I felt too tired to face the long and hot journey behind the wheel. So, I settled into the seat next to my sister, saying, *"Today, you drive! I trust you."* I fastened my seatbelt, turned the radio to my favorite station, closed my eyes, and reclined on the seat, oblivious to what was about to happen. Chiara's friend, Paola, my mother, and Billy, our old poodle who always came with us everywhere, were in the car with us.

As the car sped on the two-lane highway, and I was lost in thoughts about the previous night, Chiara overtook a fast-moving truck near Cremona. Suddenly, she swerved and lost control of the car, unable to regain it. The car began to spin, seemingly endlessly, and hurtled into the opposite lane, crashing into a truck coming from the opposite direction.

It probably lasted only a few seconds, but it felt much longer to us.

I remember desperately grabbing at whatever I could with both of my hands while in the vortex of that macabre dance, holding onto the handle above my head just as I held onto my life...

I remember my agonizing screams, and the world spinning around me, the screeching of tires on the scorching asphalt...

I remember the slow-motion images of my life unfurling before me. Then, a final thought about my love, Luca, before feeling like I was sliding into a dark and deep tunnel from which I knew that I had to return...

I remember the sharp, deafening noise of metal and shattered glass, desperate screams, and then silence, a deafening silence, and the feeling of being in a new dimension, detached from everything and everyone, finally at peace...

Then, the painful awakening from that unreal silence and the pungent smell of blood on my clothes and in my hair.

The sun visor mirror hung open in front of me. Paralyzed, I could see my reflection and the horror of my face covered in a thousand shards of glass and drenched in blood. I never imagined my blood could be so red. It was a cruel moment, being forced to look at myself, reduced to that state, unable to remove that monstrous image from my eyes. Even today, after so many years, the memory of that day is more vivid than ever in my mind; sometimes, it haunts me in the restlessness of my nights.

The first thing I saw was my face, covered in the thousand pieces of the shattered windshield, and my eyes wide open, strangely calm, staring, and distant. They seemed to belong to someone else.

The right side of my body was shattered. My collarbone, humerus, femur, and ankle all fractured upon impact with the truck. The seatbelt saved my life. Without it, I would have been thrown from the car and killed instantly.

Strangely, I felt no pain, only a blinding pulsating sensation on the right side of my body. I was perfectly clear-headed and detached. I remember firefighters, police officers, and paramedics bustling around me with anxiety and agitation. It was a strange contrast to the absolute calm that prevailed in me.

I remember asking my mother how Chiara was and how the others were. We had been lucky, miraculously, as the doctors would later tell us – everyone else escaped without a scratch, even old Billy. Everyone, except me.

My right foot was stuck in the dashboard; the firefighters struggled a lot to free me from the car, which was now reduced to a useless heap of metal.

My car had a gas tank. I heard the emergency personnel and the carabinieri say, *"She can't be alive after a crash like that. Hurry, the car could explode any minute now!"*

My mother, Chiara, and Paola were in shock. They couldn't even remember my father's phone number; they just cried and didn't know what to do. They were silent, helpless, unable to answer the questions of emergency responders. At that moment, I gathered all my remaining strength and said, *"I'm alive, and I don't want to die! I will make it. Call my father and Luca and make them come to the hospital. Here are the numbers."* I had finally returned to myself.

I then remember the searing pain that took my breath away as I was moved from the car to the stretcher. It was cold and hard and smelled of disinfectant. And the ambulance as it raced with sirens blaring to the nearest hospital in Cremona; my mother crying and holding my hand, her eyes filled with unforgettable pain, as I asked, "*Did I lose my leg*?" She couldn't answer at that moment. She didn't know. Neither did I.

What came next was unspeakable pain, both physical and psychological –lengthy surgeries, blood transfusions, being bedridden for months, and a sense of helplessness and rebellion against what had happened to me. Gradually, I got used to the sterile smell of clinics, to the long, sweltering July nights in the Cremona hospital, sleepless and aching, filled with mosquitoes instead of stars, learning to eat with my left hand, seeing the world from a horizontal position rather than standing up. But for someone as proud and independent as I am, the greatest challenge was accepting complete reliance on others for every physical need – without feeling humiliated by it.

Finally, in September, I slowly returned to life; first with two crutches, then with one, and finally, walking on my own two legs. I will never forget the thrill of standing on my own legs again after all that time.

I remember one particular beautiful fall day, and the warm, golden colors. I can see myself in that bare hospital room, thin and trembling, standing up and throwing the crutches into a corner as I started to walk, hesitant and painful, like a child facing life for the first time.

I felt a sense of victory and gratitude while tears traced paths on my sweaty face, as I took my first step without any support.

I was alone that day because I had chosen it to be. I had always been that way since I was little. I wanted to take on all of the most important and challenging moments of my life, alone. This was one of those times I had to be enough for myself.

I didn't know that my first independent steps were only the beginning. I would have to face a year of rehabilitation. Every day, I had to have massages, do exercises in the pool and gym, and undergo ultrasound therapy to regain the ability to walk. But I knew I would make it. The worst was over, and I had survived.

The immense pain of this event showed me how fortunate I was. My family and friends supported and nurtured me unwaveringly. I hadn't realized I was surrounded by so much love. Those who cared for me had kept me company on long summer days, in stifling heat, reading books to me, listening to music together, chatting or just being there, respecting my silence, putting a cool cloth to my feverish forehead, caressing me and making me feel like the most important person in the world. Slowly, I had realized that nothing had changed, except within myself, and I understood that it was all'on me to return to life.

Luca was also there by my side from the beginning. He had given me strength and courage, and he was my light in dark moments and my energy when I felt fatigued and discouraged. He showered me with love when I hated myself and made me feel beautiful and desirable when I saw a monster, turning my stricken gaze away from my scars. With infinite sweetness and patience, he always supported and protected me until I returned to life. In the end, I had hope and a desire to live again.

My accident definitely made me grow up. I became an adult. Sadly, the recklessness, bravado, and fearless spirit that had defined my youth until that moment went away. In their place, a deep sense of responsibility and an unwavering attachment to life took root. Immediately. As if it were the most natural thing in the world. I had become a new Antonella. What had happened would leave an indelible mark on my heart. It had taught me the value of life, how fleeting it was, and the importance of love.

After the accident, I learned to be fatalistic and appreciate life for the simple fact that I was alive. Perhaps that's why I finally decided to move to Kenya and intensely experience what destiny had in store for me. Coming so close to death made me realize that we create our own lives and happiness, day after day. I feel grateful, in each and every moment, for the simple fact that I am alive. Now I know that our choices, thoughts, and deepest fears and love will determine the quality of our future. It's up to us to decide what kind of life we want to live, and we are responsible for our own happiness. This was the lesson my accident taught me.

For this reason, I chose to live in Africa and love Luca, unconditionally and courageously, even though I knew it would come with a spirit of sacrifice and adjustment to a life completely different from what I had always known. I threw myself blindly into the African adventure, giving all of myself in every moment, aware that it could be my last!

Sometimes we fail to appreciate how lucky we are to live a life full of love. I know this now.

The accident changed my life and Luca's completely. Only by believing that we would lose each other forever did we understand how much we loved each other, belonged to each other, and wanted to be together. We were finally ready to experience our love and Africa. After my accident, we never parted ways again.

A New Beginning

My adventure in Africa began in 1995 alongside Luca and our wonderful friend, Giulio, who shared our hopes, feelings, dreams, and an unwavering determination to realize an incredible life project and commitment to the Maasai people: Campi ya Kanzi.

Tall, slender, with dark eyes and hair, Giulio had always been seen as being unique. Not conventionally handsome, his education, class, and graceful manners made him irresistibly captivating to many beautiful women. First and foremost, he was an *aesthete*, drawn to beauty in the people and things around him. At times, he was incredibly pessimistic and gloomy. He would retreat into his long silences, and that weighed heavy on the hearts of those around him. But he could also be brilliant and charming, excessively witty but only with those he deemed deserving of his friendship.

Though he could be impractical and inflexible, Giulio was incredibly intelligent and known for his amazing foresight, which always proved to be right! With true friends, he could be very generous and altruistic to the point of sacrificing himself if he believed it was for a just cause. He had always been Luca's best friend, from their school days, and loved him like a brother. When I met him, there was an immediate understanding and mutual affinity between us, destined to turn into deep respect and genuine affection over the years.

I trusted him implicitly. So much so that when Luca, in April 1996, decided to go to Kenya to explore the Chyulu Hills for the first time and talk to the Maasai landowners about the possibility of moving to their land, I asked Giulio to accompany him. Using his rational, objective nature and his good sense, I hoped he could convince Luca to give up the crazy idea of a life in Africa.

Giulio left, reassuring me that he would make Luca truly think and not give in to his impulses and instincts as was his habit. He was supposed to stay in Kenya for only a couple of weeks, just enough time to persuade his friend to stop dreaming and come back home to Italy. And bring him back to me.

After three months, he was still with Luca, at the foot of the Chyulu Hills, in Kenya. The two friends had already begun talking to the local Maasai and were organizing the construction of Campi ya Kanzi, the lodge that would later become our home and our new life. In short, Africa had even won over Giulio, the most rational and reflective being I knew!

From then on, Giulio would become an essential part of our African adventure and our family. He would help us build the camp with his engineering and construction knowledge. He would always stay by our side over the years with his wise and valuable advice and futuristic vision, even without his actually being in Africa.

Another key person in our African adventure was Stefano. A close friend of Giulio's and Luca's, he also came from our city and went to school with Luca as a boy. The three friends had grown up together, sharing the carefree moments of youth. Stefano was a sensitive, polite, and kind young man. He was always passionate about travel, nature, skiing, mountaineering, and photography. After completing his studies, he had decided to work with his father for a few years in the family construction business. After a while, he realized how stifled he was by the Italian way of life and decided to move to New Zealand to start anew. Like Luca, he loved his freedom more than anything and was looking for a place where he could feel fulfilled.

Stefano visited us several times during the construction of Campi ya Kanzi, always expressing enthusiasm for how we chose to live. We asked him to come to help us as a guide, taking our guests on safaris for a few months before going to New Zealand. He accepted enthusiastically. Instead of staying just a few months, he stayed for years and permanently moved to Campi ya Kanzi, playing a pivotal role in bringing our dream to life and offering unwavering support to us all. After 11 years in Africa, Stefano decided to return to Italy for personal reasons and perhaps because he was a bit homesick! I am convinced that Africa will always be an indelible memory in our friend's heart.

As for me, I was quite the city girl before moving to Africa. I lived the typical life of a young woman born and raised in a town in Lombardy – friends, parties, nightclubs, holidays, university, shopping, work... My home was a small castle in the Franciacorta region, comfortable and spacious, and I never thought I would leave it to go live in a tiny tent in the African savannah!

Until then, I had never gone camping, never wanted to, and never loved anything related to nature and outdoor activities. The first time I camped out was in Africa, in November 1995, and I must admit it was truly unforgettable.

At that time, I dreamed of living in New York or London one day, perhaps in some large ultra-modern loft, and I envisioned my future as a whirlwind of haste and life, immersed in the concrete and chaos of my beloved First World.

I was sure I wanted to spend the rest of my life with Luca, but I understood that this would mean moving to Africa and leaving behind my beloved Italy, my family, my life. Would I be able to do it?

A Choice of Love

How had Luca convinced me to be a part of his crazy African adventure? How had he won over my young heart? When had he managed to tiptoe into my soul, to the point of making me decide to leave everything and everyone behind for him?

Many years have passed since that day, but even today, I get emotional thinking about the depth, sweetness, and determination in Luca's eyes the first time he kissed me.

It was June 1988. We had met just a few days before. That evening, the moon was full, and the night smelled of magic. After a candlelit dinner, Luca took me to the rock of Manerba, an imposing rock on Lake Garda, which offered an enchanting view of the lake. It was a place that was both romantic and wild, just as Luca was in my eyes. He recited a love poem by Prévert, his gaze lingering on me like a gentle caress, and kissed me. Then he looked into my eyes for a long time and said softly, "Don't fall in love with me because I'm going to live in Kenya! If we were together, you would have to leave Italy, your family, friends, and your life here... Who knows if you would be happy in Africa?"

I should have taken him seriously and not thought that he was just a dreamer. I should have understood how important and realistic it was for him to move to Africa. But no woman, not even the most worldly, would have truly been troubled by what she had heard. In fact, when we first met, Luca was graduating with a degree in economics and already had a successful job in a business management company in Brescia. He lived in a beautiful house in the Lombardy countryside, and there was no reason to believe that he really wanted to leave Italy. So, I decided not to worry and to surrender to my great love for him, without thinking about a future in Africa.

However, I still did not know how determined he was to get what he really wanted. After 10 years of exhaustive research in Kenya, he finally found the opportunity he was looking for and said earnestly to me one day, "Remember Kenya? I found a beautiful piece of land with the opportunity to manage it with the Maasai, the landowners, and build an exclusive tented lodge for sustainable eco-tourism. We could co-exist on a wonderful, unspoiled piece of land with a quality of life we would never have here. Will you come with me? I told you I would go to Africa!"

What could I say? He actually had told me, but I had never believed him! Suddenly, my world seemed to fall apart. My plans for the future and sense of certainty crumbled before my eyes. Luca was asking me to leave everything and to go to live in Africa with the Maasai. Would I be able to do it?

Despite the fears and doubts that weighed on me during those days, one question arose insistently inside of me: "What was it about this Africa that Luca had, for so many years, dreamed of and yearned for? Could I make such a crucial decision about my future without knowing what this mysterious continent was like?"

I had to go to Kenya at least once before I could make a decision. Only then would I know what to do with my life.

After fully recovering from my accident physically and psychologically, I decided that it was time to face Africa.

So, in October 1995, I set out with Luca and my father towards this dream, or perhaps, this African "nightmare" to finally discover the Kenya I had heard so much about.

I was curious and, at the same time, intimidated by my first trip to Africa. I overcame my hesitation and fears, realizing that only by going there could I decide what to do with my future.

Like me, my father had no experience – and was reluctant – when it came to camping or adventure travel. Despite that, for my sake, he decided to overcome his fears and accompany me to find out where his defenseless little girl would end up if she decided to live in Africa. Most of all, he wanted to try to dissuade me from doing something crazy! In fact, my family had never wanted to accept my decision to go to Kenya. They had always tried to convince me to stay in Italy. Knowing me for who I was – quite the city girl – they couldn't imagine how I could be happy and fulfilled living in the African savannah, in a small tent, without any comforts or a phone, immersed in untamed nature.

Thus, at the end of October 1995, my father and I set off on our African adventure. Luca was already in Kenya and was waiting for us at Jomo Kenyatta Airport in Nairobi. We took a taxi to the small Wilson Airport where we took off for the south aboard a tiny three-seater plane. After an hour of flying over magnificent, pristine, and untamed Africa, we landed at the foot of the Chyulu Hills. We had reached paradise!

Africa Finally

And there we were, on the Kuku Ranch, before vast Kilimanjaro on the Tanzania border: my father, Luca and me.

Finally, we were in Africa, the place I had heard so much about. This is where my incredible adventure began. We spent three weeks in Kenya, in a tiny, tented camp where Campi ya Kanzi – the lodge we would go on to build with the Maasai of Kuku Ranch – now stands. It was undoubtedly an unforgettable experience for us all.

My initial reaction to Africa, unfortunately, was not a promising one. As soon as I arrived at the small camp set up for us by Luca and saw the miserable tent where I had to spend my long nights, I began to cry inconsolably. It was small, hot, and dusty, outfitted with two bare, uncomfortable, tiny cots, and a single oil lamp in the middle. It wasn't even possible to stand upright in that horrible tent!

My father slept in another tent near ours. Much to his dismay, his accommodations were equally small and uncomfortable. At night, mischievous hyenas chewed on the cords holding our tents firmly to the ground. Upon waking, we often found that half of the tent had collapsed on us!

Every day was a new battle against an Africa that was an invincible enemy for me. I felt suffocated by the infinite spaces of that untamed land. I felt like I was dying without the noises and the hustle and bustle of "my world." The silence around me seemed deafening. The lack of the familiar sounds and the multitude of people I was used to, created a sense of fear and anguish. It forced me to be alone with myself for a long time and compelled me to reflect. Unfortunately, it often wasn't a pleasant sensation.

My daily shower consisted of a plastic bucket with a shower head attached to its base, simply hanging from an acacia tree with a rope often swaying unsteadily in the evening breeze. The shower was surrounded by tall papyrus reeds, providing a bit of privacy. Of course, there was no running water or electricity in the camp. At night, only a fire under the infinite African starry sky, candles, and many oil lamps illuminated our surroundings, leaving us immersed in the darkness of the wild. Luca thought it was all very romantic, but that is not how it seemed to me.

We did laundry every morning in a plastic basin and tried not to use too much water because it came from a spring about 15 kilometers from the camp. In Nairobi, we had bought a couple of 200-liter metal drums to store the water for use at the camp. When necessary, we set out with our Land Rover pickup truck to fetch water from the spring. I was struck by how much water we typically wasted in the First World. I learned to take a shower and even wash my hair using only five liters of water. I was proud of myself for this.

We tied a rope between two acacia trees to serve as a clothesline for hanging the laundry. By evening, the warm African sun dried everything we had washed that morning. Once I permanently moved to Kenya, I would hang the many evening gowns that I had brought from Italy from that same precarious rope. My gowns were a treasure that I was not quite ready to leave behind. It broke my heart to think of them sitting in a rusty trunk outside of my tent, so once a month, I would display my Italian city garments for the African bush. The sight led to many confused, concerned expressions from the Maasai around me.

We bought an old coal iron from a hardware store in Nairobi. I remember seeing something like that for the first time when I was little with the farmers in the countryside. Even then, that strange object had intrigued me. I liked the smell it gave off, like burning wood, and the bright red color of the embers inside.

And now here I was in Africa, forced to iron using that very same object! At first, I struggled. I could never get the temperature right, and you certainly could not count on any steam for pressing. It was heavy and not very manageable and ironing a shirt perfectly without leaving wrinkles was practically impossible. I would only learn to master it with years of experience.

At sunset, at the end of a long day of work, after a hot and relaxing shower, changing clothes for dinner was a real pleasure. After a few days, that faint burnt smell that permeated my clothes, along with many other new smells and tastes, no longer bothered me. In fact, I liked it! It would soon become part of my new life in Africa.

In the camp, cooking was done on an open fire, in the shade of an acacia tree. Patrick, who belonged to the nearby Taita tribe, and was our first employee, seemingly managed to work wonders. Every morning at dawn, he lit the fire and baked fresh bread and cookies in a metal bucket cut in half, turned upside down and covered with burning embers. We even baked a chocolate soufflé in it one time!

I always wondered how Patrick knew when it was the right time to remove the bread from the oven. He said, as if it were the simplest thing in the world, that you just had to smell the bread to know when it was perfectly baked, and he was never wrong! I had tried a few times, but the bread I took out of our oven was always either undercooked or completely burnt. So, I reluctantly allowed Patrick to be the only one in charge of the kitchen, and Luca had no complaints about that.

Years later, after we had opened the lodge, we bought Aga stoves from Italy to cook for the camp guests. Patrick now knew how to cook a variety of Italian specialties, which he had learned from us, and he had become the chef of Campi ya Kanzi. Despite our insistence that he use the Aga stoves, for several months, he remained steadfast in continuing to cook over an open fire, outside the kitchen under an acacia tree, just as he did at the beginning of our African adventure.

We had also bought a small gas refrigerator in Nairobi. I was impressed by how well it cooled. It was a great luxury to have, and it allowed us to keep, in addition to many other foods, excellent Kenyan beer, White Cap, salami, and parmesan cheese brought from

Italy, which we all loved. Every two weeks or so, we had to drive for two hours on a dirt road to reach Mtito Andei, the only village where we could buy a new gas cylinder for our precious refrigerator. But we did it with pleasure. In fact, it was an opportunity to go on safari and see new animals.

Our life in those long-ago days was simple, made up of small gestures and nuances; we felt new emotions every day. Together, we rediscovered our love for the little things, which had, at that point, lost their meaning in the world that I came from. And yet, thinking back, I was simply happy and serene then, without my even realizing it. My wishes were fulfilled, and I was enough for myself.

There was no telephone or television in our small camp. Communication with the outside world was provided by an old HF radio, a crackling object that almost never worked. Initially, I was bothered by the idea that no one knew where I was and that I couldn't communicate with a living soul. I was always used to having my life under control, to being able to handle situations with ease. Now I found myself in a new, hostile, wild world that I couldn't understand or control. I was at its mercy, and it terrified me.

Everything around me was harsh, acrid, and tough. The thorns of the acacias covered my skin with scratches, and the flies gave me no peace. The dust was fine, like talcum powder and made its way into every part of my body – under the nails of my hands and feet, in my ears, nose, and hair. I always felt dirty, even though I washed every day.

The hot, suffocating climate made me nauseous. I was exhausted, I felt like I was slowly dying. I felt drained by Africa, more and more every day. And the fear of wild animals, insects, and snakes, gripped my stomach and led to sleepless nights.

On my first night in Kenya, when we arrived at the tented camp, I expected Luca to stay there with my father and me. However, at sunset, he had decided to go see the elephants drinking water from the lake in the moonlight, about a three-hour drive from our camp. He had asked my father and I to go with him; but we were exhausted after the long journey from Italy and decided to stay at the camp and go to bed early. I thought Luca would stay with us, given the discomfort and fear that my father and I felt in that situation. But he was not fazed and left for the lake, without any second thoughts, leaving us alone to face wild Africa!

Only now do I understand that my refusal to go with him was probably a sign of my desire to test myself and challenge my anxieties. I wanted to face that foreign land alone and immediately overcome my fears. I couldn't bear the idea of succumbing to that wild, yet fascinating country. I wanted to try to adapt to it without regrets, no matter what decision I made.

But my father and I did not know of the darkness looming in the bush at night, even with a full moon. The usual city sounds we were accustomed to were replaced by unfamiliar cries from wild animals, echoing in the silent darkness.

So, to warm our hearts, gather our courage, and forget about the nearby cries of lions and hyenas, we started to drink some of the wine we had brought from Italy.

When Luca returned the following morning, he found us completely drunk but blissfully asleep by the fire. It had been a long night, our first night in the bush. But we had bravely survived, with a little help!

Longido

The morning after our drunken adventure, we decided to make ourselves a cup of tea to recover. When we opened the sugar bowl, we were confused to see no sugar in it. Instead, we found an army of ants that had eaten all the sugar! So, we unhappily sipped our sugarless tea.

It was a beautiful day, and an intense, stifling heat began to rise from the sun-cracked, red earth. It is impossible to describe to those who have never been to Africa. It is a heat I would later get used to over the years, but in those early days, it seemed unbearable.

After breakfast, we decided to take a walk to the top of a rock that the Maasai called *Longido* (granite in Maa, the Maasai language).

We were accompanied by a Maasai guide. Nkalami was handsome and proud in his fiery red *shuka*, draped over one shoulder in the manner of the toga worn by ancient Romans. Tied around his waist was a belt of beads from which hung a *simi*, a long dagger kept in a dark red buffalo leather sheath. While waiting for us to finish our sugarless tea, he comfortably rested on one leg, leaning on a slender ebony stick, in the typical Maasai position. One leg was straight and rigid, and his bent leg rested effortlessly on the other. Focused and silent, it seemed like he was in a yoga pose.

He gazed at the horizon, eyes half-closed, as if meditating. In one hand, he held his spear, and in the other, his *rungu,* a short olive wood stick, a symbol of power that Maasai village chiefs always carry. He looked into the distance, silent and deep in thought, toward a point unknown and unfamiliar to us. It was a splendid sight: fascinating and primitive. He was a *moran*, a warrior, and he had a great sense of dignity and pride.

As we walked, I enjoyed observing him. He had long, slender legs, like the limbs of a flamingo. As he walked elegantly ahead of us in the tall grass, he seemed to glide over the ground, light as a feather carried by the morning breeze. My gaze fell onto his well-shaped and strong shoulders. His long, braided hair, colored with red ochre, rhythmically waving down his back, was crossed by colorful coral-beaded necklaces that stood out beautifully against the ebony of his skin.

He made his way slowly through the thicket, his long spear resting on one shoulder, and sniffed the air, seeking the trace of a familiar smell carried by the wind, scanning the horizon with impenetrable eyes, searching for animals that were invisible to us. We followed him in silence, with the blind devotion meant for the most magnanimous and great Gods. We were aware of the fact that our safety rested completely in his hands.

Longido is an ancient granite rock. It erupted from the bowels of the Earth over 20 million years ago. From its summit, one can take in wonderful views of the infinite beauty of Africa. It was formed by various dark boulders and a perfectly vertical rock wall overhanging the immense plains facing Kilimanjaro. At sunset, it was tinted with pink, red, and orange, and its iron-grey veins stood out like ancient hieroglyphs.

My father and I were not very keen on the idea of walking in the savannah, but we let ourselves be won over by Luca's enthusiasm. We were scared and uncomfortable in that environment that was so alien to us. Our senses were on high alert, incredibly tensed up and ready to pick up on any sign of the presence of wild animals nearby. Luca carried no rifle, and Nkalami had only a spear with him. Nothing more.

I instinctively trusted Nkalami, but not completely. I would learn to do so slowly, only after spending a long time in Africa. After all, that had always been my nature: cautious and rational to a fault and distrustful of everything and everyone, especially if they were new to me.

During the walk, we encountered a leopard. I realize now how incredibly lucky we were. The leopard is a very rare animal to see, especially during the day. It seems that this was Africa's way of welcoming me! But at that time, unfortunately, I did not understand or appreciate it. Luca was thrilled and followed the leopard with Nkalami, climbing a hill in front of us, leaving my father and me defenseless at the foot, back-to-back, terrified, and waiting for an enormous beast to tear us apart! They were back in a few minutes, but it seemed like we waited an eternity. I remember never being happier to see someone than I was that morning, seeing Luca and Nkalami reappear on the horizon, descending the hill, and coming toward us.

We returned to the camp at sunset, safe and sound, but the emotion of that walk would stay with us for a long time.

At the end of those first days in our camp, Africa, with its inhabitants and wild animals, had already managed to leave an indelible mark on my heart. I believe the Maasai were equally struck by their first impressions of me.

I had been so terrified by those ants in the sugar pot that one morning, they found me tormenting a poor blue lizard that had the bad idea to climb to the top of my tent to sunbathe with an acacia branch full of thorns. It may as well have been a monster to me, so I started to hit it angrily and ferociously until it fell to the ground lifeless. I was unaware of the fact that I had an audience – five wide-eyed Maasai and Luca who were bewildered by my behavior. They stood there shocked and confused as a hysterical city girl with a thorny stick in hand beat a lizard as if it were a lioness!

Even today, we laugh at those first days in the African savannah. They say it was already clear that I had a strong character! Even now, I feel guilty for having killed that poor lizard; today, lizards are my children's favorite playmates.

When Maasai men marry, they must pay a dowry to their future in-laws in the form of cows and goats. During those early days at the camp under Kilimanjaro, we joked with the three Maasai who were our guides. One day, Luca asked Nkalami: “If Antonella were a Maasai girl, how many cows would it cost to have her as his bride?” The warrior thought for a few minutes, then exclaimed, “Ten!” I felt very proud of myself. Usually, the average dowry paid for a beautiful Maasai bride was around five or six cows. Ten was an exorbitant amount! Immediately after, however, Nkalami added, “However, Antonella’s father would have to pay her future husband, not the other way around! No one would want Antonella as a wife. She would be a *bibi kali sana!* (a very difficult and demanding wife). She never wants to go gather firewood or fetch water from the spring. She gives orders to everyone and never obeys....” I realized then that the Maasai sense of humor was one of their best qualities.

Days of anxiety, stifling heat, and scorching sun followed on lazily, but, to my surprise, each day seemed less harsh than the one before. I was beginning to acclimate to Africa. I was beginning to love it, without even realizing it. Then, on the last night at the camp before returning to Italy, the rains began.

Even then, I was not prepared for what was about to happen. I remember thunder and lightning of such strength and intensity that it shook the Earth and lit up the dark night. We were flooded by a river of water, which flowed endlessly, taking over everything in its path. The wind howled outside of our shelter, violently shaking the tiny tent. I was terribly frightened, feeling lost and at the mercy of this wild Africa, and totally powerless in the face of this spectacle of nature. A wall of water surrounded us, threatening and impetuous. I was breathless from discomfort and fear; I felt like I was drowning, even without being in the water.

Then, suddenly, everything quieted down as quickly as it had begun. In the ashen sky, a cobalt blue tear opened, embroidered with a thousand sparkling stars and a slice of silver moon. The crickets began to sing again, and the frogs croaked peacefully. We heard the cry of a hyena in the distance.

It was as if nothing had happened. It was as if everything slowly exhaled and returned to the usual ancient rhythms. Africa was once again a paradise with its absolute beauty and tranquility. And, finally, my heart, too, returned to peace and calm. I rested my head on the pillow, let myself be lulled by the song of the crickets, and peacefully fell asleep.

I felt happy, perhaps for the first time in my life, without realizing it and without understanding why.

We left the next day, reluctantly, in a small three-seat plane to return to Nairobi and, from there, return to Italy. A strange restlessness crept inside of me. The time had come for me to decide if I could live in Kenya and if I wanted to.

Back to Civilization: Nostalgia for Africa

After my first brief stay in Kenya in October 1995 with Luca and my father, I returned to the "civilized world" of my Italy. But I could not understand if I was truly happy.

I began incessantly ruminating on what I had experienced in Africa during that brief but intense stay in the tented camp at the foot of Kilimanjaro. I realized that I felt nostalgia for that land and its people, who stirred strong and incredibly genuine emotions in me.

Africa was in me; I know it now. It had slowly made its way in without me noticing, into my mind, my body, and, more dangerously, into my soul. In my eyes, the infinite African sky; in my mind and heart, the peace, serenity, absolute freedom, and excitement I experienced in that distant land. In my soul, the desire that had now become an urgent need to get to know Africans, especially the Maasai, and to share my life with them. A sense of restlessness and fear rose in the darkest and most private corner of myself as I understood, against my will, that what I had in Italy was not enough for me, nor would it ever be. Something so far from being true and authentic could never compete with everything Africa was.

I seriously began to consider the possibility of leaving everything to live with Luca in Kenya. I had only one question that I needed to answer with utmost honesty: "What do you really want from life?" This is possibly the most difficult question I have ever had to answer!

It took me six long, agonizing months to decide. During this period, I never shared any of my innermost doubts and fears with anyone. Especially Luca had to stay away from my turbulent path, allowing me the true freedom to make my own decision, whatever it might be.

Then, slowly, a certainty grew within me. In the midst of the countless questions that rocked me like stormy sea waves, one thought stood out clearly like a lighthouse in the night, solid and shining bright. During the few days I spent in Kenya with Luca, I realized how much that land meant to him. Luca was completely different when immersed in the wild, in contact with the Maasai, embraced by Africa's strong arms. In his eyes, I saw an inner peace and profound calm that I had never seen in him before. In Africa, Luca was finally at home.

I could not and did not want to ignore any of this, and spiraling thoughts raced through my mind during that period: "I want to spend my life with Luca. It doesn't matter where, as long as we are together. I can't ask him to give up Africa. That mysterious land is not a whim; it is in his blood, in his soul. It is part of him. I could never accept being with a man who was forced to sacrifice a part of himself and his dreams because of my fears... if I love him with true and authentic love, I cannot ask him not to be free to live out what is inside himself... if he were no longer free, he would not be himself anymore, and I couldn't love him as I do now."

I have never regretted choosing Luca and, therefore, Africa. I am the first to admit that, at times, I realize that this difficult choice has put me and my nature to the test. But thd joy I feel every day in seeing myself and the man I love successful in what we do with the Maasai, makes me feel lucky and privileged. It makes me feel proud of us both. It makes me want to teach my children, as the Maasai teach me every day, the values of love, freedom, empathy, and truth. The importance of substance, not form.

This is what I learn from the Maasai every day: respect for the elderly and protection of women, children, and the sick; the importance of living today and the awareness of the fleeting nature of life; the values of unity and solidarity, patience and discretion.

These values are fundamental to the Maasai community. They seem to be almost lost now in the "First World."

The Maasai respect and honor their elders. Unlike the world where I was born, in Kenya, old is synonymous with wise and does not mean useless. Young people listen to the *wazee* (the village elders) and seek their advice for any problem or issue that needs to be resolved.

They protect and defend their children and women. The Maasai women play a role that might seem submissive to many. From the beginning of time, it is the women who have dealt with the household budget, raised the children, and built the huts. Life in the *boma* (Maasai village) revolves around the women, flowing peacefully and unchanged for generations.

The Maasai take care of their sick with a patience and dedication that I have rarely seen in the modern hospitals of the western world. They live only for today with an intensity and depth, not thinking about the past and not worrying about the future. Isn't that how life should be lived? What is the point of living with regret or dwelling on future plans if that means missing out on the truth and intensity of the present?

The Maasai help each other. They do not envy things that belong to others because they do not know the concept of private property. For them, everything is for everyone, or rather, it is for those who need it at that moment. They are a large family, and we immediately felt like we were an integral part of it.

For us, accustomed to western individualism and selfishness, used to consumerism and the rush-rush nature of the modern world, learning the ways of the Maasai was strange at first. We had to change our priorities, and it was a bit like going against our nature. Now, we are happy and grateful to the Maasai for having retaught us the values of life and for welcoming us into their family as fathers and mothers, as brothers and sisters, as sons and daughters.

Learning the Ways of Africa

It was July of 1996, after months of emotional turmoil, that I decided to follow Luca and move to Africa permanently.

I boarded the first plane to Nairobi and suddenly found myself catapulted into a new world made up of scents and colors so intense they made my head spin. The first thing that struck me as I exited the airport in Kenya was the multitude of people wrapped in vibrant shades with eyes so deep, so gentle and so kind.

Not once did I feel threatened or unsafe among Africans. Not once did I feel alone, I was always a part of the community and in harmony with everything around me.

One of the official languages of Kenya is English, but not all Kenyans have gone to school and can speak it. Kiswahili, the other official language, is spoken by almost everyone. I adore it for its musicality. It is a sort of *lingua franca* that came about on the coast during the slave trade and spread throughout Kenya, to Tanzania, and to parts of Uganda. It is the "*Esperanto*" of Africa.

Unlike Luca, I did not speak English when I got to Kenya. I had studied it in school, but I had not spoken it for 10 years, so I struggled to bring it back to mind. I didn't even know about Kiswahili before coming to Africa.

Thus, it was not easy for me to communicate with Africans in the beginning. Nevertheless, I managed to communicate at once with the Maasai at our camp in the Chyulu Hills, the "Green Hills of Africa" according to Hemingway. The property extended for more than 1,000 square kilometers. The only ones living in that still unspoiled, untamed corner of Africa were us and 3,000 Maasai.

We felt fortunate and privileged for this. We decided to build a tented lodge, *Campi ya Kanzi* (camp of the hidden treasure, in Kiswahili). It was to be small and exclusive, ecological, and respectful of the environment and the surrounding animals. We wanted to promote eco-tourism and to guarantee the preservation of that piece of Africa and its proud, indigenous inhabitants. And that is what we did with the local Maasai.

It was a great adventure for everyone; the realization of an incredible dream from nothing. We had to build everything from scratch, even the roads. In short, it was an undertaking worthy of true pioneers: and to think that I had never felt adventurous before!

Most of the Maasai who helped had not attended school, and only a few had seen a white person before. But this made everything even more beautiful and compelling.

Those who learned to be masons and electricians at the beginning of this venture have now become waiters, cooks, drivers, and guides. The greatest satisfaction is certainly the fact that Campi ya Kanzi is not only our home but, above all, it is the home of the Maasai who work there and who have now established a relationship of kinship with us and a great affinity with the camp's guests.

At the beginning, I needed to communicate mainly with the Maasai women, and I was truly curious about them because they were so different from me. I had to teach them to clean the tents, make beds, wash and iron, in short, all sorts of things that would make them perfect "homemakers!" There was also a lot to teach the cooks when it came to preparing the perfect *pasta al dente* or *tiramisu*, especially considering we could not communicate in the same language.

I showed them everything, and they learned so quickly. An Italian friend, Monica, came to visit me once and taught me how to make a chocolate cake. She wrote the recipe on a paper napkin that I still have and need to reread to make the same cake after so many years.

Nareio, one of the Maasai women I had just hired at the time (who could neither read nor write) observed everything with great attention. The day after my friend left, Nareio called me into the kitchen and with a brilliant smile took out of the Aga oven an amazing chocolate cake, undoubtedly better than the one I had baked the day before following the recipe. She had seen Monica as she was making the cake and had quickly absorbed the necessary information to make the same dessert in no time. She didn't need to read the recipe to make it, unlike me, who still needs it today. Nareio still makes the best chocolate cake and still doesn't know how to read!

I am always amazed at how quickly the Maasai learn how to do the things that interest and appeal to them. They do not need to write anything down; they remember everything, perhaps because theirs is a spoken, not written language. What they know is passed down by word of mouth by the village elders, who immediately begin telling the children about the most important and famous achievements and adventures of their proud and courageous ancestors. So, they get accustomed to remembering, without the need to write anything down.

When I think back to those first days when we were building the camp, I see myself sitting on spools of electrical cables in the middle of the savannah, intent on trying to converse amiably with old Maasai women in front of a steaming cup of tea. They talked to me about their husbands, their children, and their huts. They talked to me about their lives. And I, inexplicably and mysteriously, understood, or rather, picked up on what they wanted to tell me.

Thinking back, I find this extraordinary! I longed to communicate with them so much that I could do so even without speaking much Maa or Kiswahili. In just six months, living with the Maasai, I learned Kiswahili perfectly. It is a language that still fascinates me, and I love to speak it. Its musicality and sweetness make it a special language and very dear to me.

As the camp was being built, my active participation was limited, as I knew nothing about electrical systems, plumbing, level canes, cement, and lime. For someone like me, practical, energetic, competitive, and wanting to have a well-defined role, this period was frustrating. And this explains my desire to converse with the Maasai women, how I learned Kiswahili so quickly, the creation of such an intense bond with Africans, and, especially, with the Maasai.

I am almost certain that the Africans' desire to understand others made up for my lack of knowledge of their language. So much so that I rarely felt the frustration of not being understood by them. They are so kind and polite that they will never embarrass you if you cannot express yourself properly in their language.

At times, we may suffer from frustration coming from the "First World" and having been raised differently. We like to have all the answers to our questions at once. Before getting deep in conversation, the Maasai engage in a series of customary pleasantries. Failing to observe these rituals marks a person as impolite, coarse, and ultimately unworthy of serious consideration. These pleasantries could be about rain, family health, the well-being of wives and children, and the number of cows and goats they have.

The Maasai will hardly ever contradict you or tell you that they didn't understand what you have said. They will barely ever refuse your request, even if they know it will be impossible to achieve. They will instinctively sense the response you want to hear, and that is what they will give you. It is not that they are liars. This is just their way of respecting and not offending you. Unfortunately, for many westerners who see their responses as lies, this behavior is unacceptable and insulting. It is actually a form of courtesy.

At first, I had a hard time accepting that the Maasai, or rather, most Africans were like this. For me, like for most of those who come from the hectic western way of life, it was essential to know and decide upon everything at once. Then I began to appreciate their sensitivity toward me and others. Today, when I see outsiders being impatient and rude with Africans, I am annoyed and hurt.

Who knows how many times I have inadvertently annoyed and hurt Africans? I truly have my regrets about it.

Maasai Women

They resemble slender reeds in the breeze. Delicate and lean, elegant and sinuous, so fragile that they could break at any moment: yet they do not bend, do not break, even in the most impetuous wind.

They are strong and stubborn, willful and proud, joyous and talkative, noble and wise, yet untamed: these are the Maasai women.

Their bodies, slim yet strong, curve under the immense weight of bundles of firewood and buckets of water to be carried from the spring to their homes. Yet, every day, they hang in there with tenacity, as they have done for thousands of years, against the mistreatment and abuse sometimes inflicted on them by their men, fathers and husbands, life companions and forever their "masters."

Their perfectly shaved beautiful heads shine in the African sun. They wear a rainbow of beaded, round necklaces in bold, intense hues, in various sizes that almost cover their shoulders around their long and slender necks, like *gerenuks*[1]. Long, flashy leather earrings, decorated with colorful beads – striking yet graceful – dangle from the lobes of their ears.

Bracelets and thin circles of dried tinted seeds, ostrich eggs, and shells decorate their delicate wrists; beaded belts and cowhide cinch their waists and drape elegantly over their red and blue *shukas*.

They have broad foreheads and high cheekbones, adorned with elegant tattoos. And delightfully sharp noses, small ears with pierced lobes, slightly pointed chins, and dark intense eyes. White, beautiful teeth, like rows of rare and precious pearls, light up their faces suddenly when they smile.

1 A giraffe-antelope, which gets its name thanks to its very long slender neck.

They feel beautiful and regal, proud to be Maasai. And they are.

Cloths of blue, red, and white fabric elegantly cover their slender and naturally sensuous bodies, and around their ankles, they wear beaded bands which make a jingling sound in the African dust as they walk barefoot, tirelessly covering infinite distances.

When I visit them in the village, they gather around me and begin to sing and dance, slowly and rhythmically. If I close my eyes and let myself be carried away by their songs, I begin to move my body with the rhythm until I feel like I am a part of them. My voice joins theirs in a single song, in a single melody, in a single breath. Then, I feel as if we are so close even though we are so different. Then, I feel like one of them.

Lupembe, the leader of the Maasai women of Iltilal, the village closest to Campi ya Kanzi, which is a two-hour walk away, always receives me with joy. She comes to greet me, majestic in all her splendor of colored jewelry, and smiles at me, mischievously. She has a beautiful face, an open smile, and the direct and sincere expression of someone who is not afraid to say what they think. She takes me by the arm and begins to tell me about the latest news from the village. We respect each other; we share a great desire to communicate and make the conditions of the Maasai women better and more tolerable.

The children, shy and bashful, approach me, bowing their heads so I can place my hand on them, a sign of respect. This is their greeting, and it always embarrasses me, but I know that if I refused to caress them, they would be offended. Then they smile at me, take me by the hand, and start calling me by the Maasai name they gave me when I arrived in Africa: *Naseryan*, the one who helps. It is nice to be called that. It gives meaning to my life and makes me proud of what I do.

From the time they are still children, Maasai women always carry a newborn – their child or their sibling – tied on a bundle. Rarely have I seen a Maasai woman without a little one in her arms or on her back. These proud people adore children; perhaps because, before the medical care now offered by the government, infant mortality was extremely high. The Maasai say it is better to be a man without cows but with a child than the opposite. Considering the enormous importance that cows have for the Maasai people, this statement shows the love they have for their little ones.

Maasai history tells of a powerful, benevolent God, called *Enkai*, who created the Maasai first and then gave them all the cows in the world. The Maasai are known for the cattle raids they have carried out since the beginning of their history.

Nomadic herders, of Nilotic origin, the Maasai moved from North Africa with their herds to settle in the plains of Kenya and Tanzania, where they are still today. During their movements, they were feared and invincible warriors, and conquered all the other tribes they encountered, capturing their cows and women. For the Maasai, however, taking the cattle of others was not theft. It was just reclaiming what *Enkai* had given them at the beginning of time, namely, all the cows in the world!

When the Maasai have children, it is considered a true blessing. They love them and play with them for a long time, patiently and cheerfully. Yet they do not hesitate to be authoritative and strict when needed, teaching their children respect and good manners.

The role of Maasai women in their society seems limited. But they play a fundamental role in the lives of their families and are proud to do so, although often with difficulty. They manage the family budget and are the keepers of the most private secrets of women. The grandmothers will teach their granddaughters, from an early age, how to care for goats and cows and, in the same way, how to be perfect wives, mothers, and homemakers. They are the only ones who have a say in managing the children and the household. They decide on the education of their children and prepare infusions and herbal teas with medicinal herbs and roots found in the forest to treat the sick. They expertly shave the heads of their proud sons before the important circumcision ceremony that will make them *morans*, warriors. With sweetness and patience, they dress and adorn their beloved daughters before the wedding ceremony that will take them away from home, courageously hiding the sadness in their hearts, knowing they will lose them forever after that day.

Undoubtedly, the quiet rhythm of life in the *boma* and the balance found in the families has everything to do with the intelligence and strength of the women. The men will always protect them. Although they can be intrusive and overbearing at times, they will always take care of them and provide for them as best they can.

The extraordinary strength and the legendary attachment to the traditions of the Maasai people is mostly thanks to the women, who with infinite patience and tenacity, stand by their men and support them in every decision.

When I arrived in Kenya, I decided to hire three Maasai women to cut dry grass to be used for the construction of Campi ya Kanzi's roofs. It was difficult to convince the village elders. They did not want to give permission for their women to go far, to have a life separate from theirs, and to be outside of their control. But I insisted and managed to get what I wanted.

The Maasai women – Serah, Salayior, and Nareio – initially looked at me with a sense of distrust, but they also had a great deal of curiosity about me. It was not easy for me to approach them initially, and they deliberately kept their distance. Yet, I instinctively felt that they wanted to build a relationship with me. It was just a matter of waiting for the right moment to earn their trust.

When the camp was built, I decided to employ the same women to help in the kitchen, clean the tents, do laundry, and iron. It had been almost two years since I had come to live among the Maasai, when one afternoon, the eldest, Serah, invited me to tea. I understood that the much-awaited moment had arrived. Happy and excited, I followed her inside their hut.

They had lit the fire, prepared the tea, and we started talking. I remember it as if it were yesterday. I did not speak Maa; they did not speak Kiswahili or Italian. Yet, we managed to communicate, and I tried to understand their lives. It was the beginning of something great and moving: the birth of a new Antonella.

After realizing the difficult situation the Maasai women were in, it became a kind of mission for me to employ them and give them the opportunity to earn money at the end of each month, allowing them to be independent of their men. It was also my desire to instill within them the awareness of being individuals with the right to be themselves and to aspire to the life they wanted.

Today, I employ 20 Maasai women at the camp. I like to think that I have developed a special relationship with each of them, and I have helped them grow, no longer afraid to voice their needs and worries.

They trust me, and they are my friends. They confide in me when they have a family problem; if their husband beats them or if their children are sick and need to be taken to the hospital, I am the first person they ask for help. They turn to me for a loan to send their children to school or to rebuild their hut after a fire or after being destroyed by an attack by a hungry lion. They console me when I am sad and come to visit me if I am sick. They play, dance, and sing with my children, and they make us feel their deep affection every day.

I am eternally full of profound admiration for these women so challenged by life, and I try to learn from them in terms of the spirit of sacrifice, tenacity, and strength needed to live every day.

I feel solidarity in the hard battle of their everyday lives and try to make their living conditions more dignified and less harsh.

I hope that my daughter Lucrezia can, one day in the not too distant future, continue this project of mine and feel the same bond I have with my dear Maasai friends: I would be proud of her, and I know that it would be a great privilege to do so.

Guardian Angels

In our early days in Kenya, during the construction of Campi ya Kanzi, we often had to drive from the camp to Nairobi and back to buy the building materials and food supplies we needed.

Usually, once or twice a week, we found ourselves, much to our dismay, on the only paved road connecting Nairobi to Mombasa on the southern coast of Kenya. Using the word "paved" is a stretch! Actually, the asphalt was full of large dangerous potholes (especially after rainy season) that left no room for distractions while driving and often caused severe damage to our cars.

The perilous and critical conditions of the road were unimaginable. It is ranked as one of the roads with the highest number of fatalities worldwide each year.

It was a narrow, bumpy road, with just two lanes, one for each direction. The problem is that most of Kenyan drivers, perhaps because they are naturally reckless and lack a sense of limitations on the road, always drive at full throttle. They only brake when necessary, meaning right before they crash! It doesn't matter if going downhill or uphill, if there are curves or bridges to cross; the African driver will always go at the maximum speed allowed by their vehicle. Moreover, to avoid the frequent holes along the sides, they always drive in the middle of the road, never in their own lane. Therefore, if we wanted to avoid being literally thrown off the road, we had to stay in the middle of the road like them, until the very last minute, to force them to move over and give us space. It is one thing to play chicken as a child with your friends, and quite another when your opponent is a huge truck threatening to send you rolling off the side of a cliff!

Thus, every journey on that terrible road was neither relaxing nor pleasant. We would arrive at our destination exhausted, physically and psychologically, and would thank God for being safe and sound. Especially for me, already greatly put to the test by the horrible experience of my car accident in Italy, every moment spent on that road was true suffering.

There were often remnants of an accident on that road. Overturned trucks, cars, and *matatus* – typical 12-seat minibuses in Kenya that carry up to 20 people, crammed together – literally demolished, huge buses without a wheel or with a completely bent axle. And often, there was at least one dead person, if not more, lying motionless by the roadside, surrounded by a curious crowd of onlookers. A macabre and gruesome spectacle that we had to get accustomed to over time.

To go to our camp from Nairobi, we would drive about 230 kilometers to Mtito Andei; there, we would enter Tsavo National Park, crossing westward towards Amboseli, to finally reach the camp after 70 kilometers of dirt road.

It was a long and tiring journey, usually taking us about six hours. We always set off early in the morning to have the whole day ahead of us in case we got a flat tire (which happened regularly) or had any other mishaps.

And yet, I remember those trips with joy in my heart and a touch of nostalgia. The surrounding landscape, as we drove, was fascinating and constantly changing. It offered a glimpse into the real Africa and the people who inhabited it.

Today we have an airplane and fly to Nairobi whenever we need to. The flight only takes us an hour and is much more pleasant and relaxing than our usual drive. Luca is now a pilot, and often flies low over a wonderful, still primordial and unspoiled Africa, looking for wild animals below us. Although I am not entirely comfortable in the air, those flights are always thrilling and fill my heart with light and energy.

But the intense sensations experienced while driving on the road, the vibrant colors of Africa, its intoxicating scents, its barefoot and smiling people selling wild honey and squawking chickens by the bumpy roads... all this I no longer see, and I must admit, I miss it more than I would have imagined.

Leaving Nairobi for the camp, we would pass through the city center. It was a large, chaotic metropolis, full of contrasts and contradictions, of rich and poor, yet very alluring. We would enter the industrial area, dotted with colorful shacks with incandescent tin roofs shining in the sun, next to ultra-modern factories and businesses.

Often, during periods of drought, we would encounter Maasai shepherds, dressed only in their fiery red *shukas* and tire sandals, grazing their herds serenely and unperturbed, even in the city's lush gardens and parks. It was a sight that always brought me joy and reminded me that, despite the west's attempts to tame Africa, this splendid continent remained wild!

Nairobi means fresh water in Maa. The Maasai shepherds used to come during the dry season but stopped when the British built the railway at the end of the 19th century. The main construction camp for the railway was established where there was water, and so Nairobi was created just over 100 years ago.

Observing the Maasai shepherds with their cattle crossing the streets of Nairobi, I think of their great-grandfathers. It makes me reflect on the revolution and transformation brought to the area, which may not have been as beneficial as one may think. Who knows if the Maasai, deep down, are aware of it and truly want to be part of the civilization that invaded them?

Upon leaving Nairobi, we would enter the Athi River area (call this because the Athi River runs through it), characterized by its ostrich farms and large, colorful greenhouse flower crops. Kenya exports many flowers to Europe.

We would continue for about 100 kilometers before reaching a village, Sultan Hammud, named after a great sultan of Zanzibar, who stopped there to rest for a night a century ago.

In Makindu, not far from Sultan Hammud, there is a beautiful Sikh Indian Temple, which is still a pilgrimage destination for Hindus and non-Hindus. For those who want to stop and rest during their journey, the temple's door is always open, regardless of the traveler's religion or race. They provide food and drink, a clean bed, and blankets, without asking for money or anything in return, as a way of honoring the sacred duty of offering hospitality. It is said, even today, that if you stop at the Sikh Temple in Makindu and say a prayer, leaving an offering (even a tiny one), your wish will be granted. One day, not long ago, a dear friend of mine, Sarita, stopped at the temple and prayed for us. We were going through a challenging time at the camp for various reasons. What Sarita asked for was granted. Immediately. I like to believe that the spirits of the Gods of the Makindu Temple are our friends and helped us in a moment of difficulty.

There, near the temple, it was possible to catch a glimpse of the red dirt road of Emali and the old railway that still connects Nairobi to Mombasa. On clear, crisp days, you could also see the snow-capped peak of Kilimanjaro. The sight of that immense mountain always stirred great emotion in me. From that moment on, I would start to feel at home.

Leaving Makindu behind us, the vegetation changed. The imposing baobabs made their appearance. These enormous African trees are characterized by a massive, wrinkled trunk and many twisted branches, mostly without flowers or leaves for much of the year. The huge *kopjes* (dark granite boulders) also became part of the landscape, and the climate grew increasingly dry and hot.

Finally, we would arrive at a small clearing of yellow fever acacias, which are easily recognizable by their yellow trunks and tender green leaves on their branches. We knew that these trees grew near water, and indeed, at Hunter's Lodge, we would find a fresh little stream that flowed for most of the year. The lodge was built in the 1950s by a famous English pioneer who arrived in Kenya at the beginning of the century. Tables and chairs were set up in the shade of acacias, on a green and soft grassy carpet, by a pond where large tilapia fish swam. We often stopped at Hunter's Lodge to break our safari and eat something. It was pleasant and relaxing and made the long journey home more bearable.

About 70 kilometers from Hunter's Lodge, a dilapidated and decrepit wooden sign on the roadside announced that we had reached Mtito Andei, meaning "forest of vultures" in the Kamba language.

It reminded me of a Wild West village from the John Wayne movies I loved as a child, uninhabited and bare, abandoned after horse-riding bandits robbed it! The largest and most important buildings stood on the main road. It was the last outpost of humanity before we got into the real bush and continued to the camp. There was a strange, almost unreal atmosphere. It seemed desolate, crumbling, and lifeless; yet, inside, it teemed with activity, energy, and people. It was fascinating.

The village was a cluster of small houses, with walls painted in vibrant but peeling hues with roofs made of straw or tin. The roads were dirt tracks, dusty and riddled with potholes, with stones scattered everywhere. It was a cheerful whirlwind of colors and activity. The people, who were always smiling, busied themselves with a thousand different things.

There was the fruit and vegetable market, full of bright hues and the shouts of street vendors. There was the butcher, who hung huge, foul-smelling pieces of raw meat outside his shop. As soon as the sun was high in the sky and the heat began to rise, flies would swarm by the thousands, covering the meat until it was completely black. There was the tailor/cobbler, who, with his old pedal-powered Singer sewing machine, sat outside his shop, and was miraculously able to repair any type of clothing or shoes. And there was the inevitable Coca-Cola vendor, selling other soft drinks as well. I could never understand how they were always perfectly cold!

Incredibly, there was also a functioning post office, where we had our mailbox and collected all the mail for our camp; and a shabby, dirty bank with just one employee, without a phone or computer, but equipped with an obsolete calculator and old ledgers in which he would record the daily inflow and outflow of money with a pencil. There was also a ceiling fan, intended to provide relief from the sweltering heat for customers. Naturally, it had not worked in years, but no one dared remove it!

I remember one day I went to ask for information about my credit card, and the clerk's eyes widened in astonishment, saying he had never seen one in his life before. I still had much to learn about the efficiency and functionality of banks in Africa.

Completing the picture of this strange little town was a Total gas station, with four pumps, and typically only one worked, and a bar, the Paradise Hotel. Here one could stop in to drink a Coca-Cola or eat a *chapati*, a type of flatbread originally from India, but now quite common throughout Kenya, while waiting for the car to be refueled. Filling up the tank could take up to 30 minutes; African time, you know...

There was one thing that always made me smile whenever we sat at the Paradise Hotel. I would look up at the ceiling and see dozens of enormous black bats, sleeping peacefully, hanging upside down from the ceiling beams, gently swaying in the day's breeze. The first few times, that sight killed my appetite; but eventually, hunger and fatigue took over, and I was able to eat peacefully under the bats, even finding amusement in watching them!

Of the countless journeys we took, one remains etched in my heart, unforgettable because it was magical.

One day, while driving near Sultan Hammud, suddenly, two magnificent greater kudus appeared from the bush, right at the edge of the road: a majestic male with beautiful horns, and a timid female. Greater kudus are very rare and stunning antelopes. Seeing them is really difficult, and to be able to observe them in such an environment and context was incredible! It was the first time since I had arrived in Africa that I was so lucky.

So, we pulled our Land Rover to the side of the road and remained there, enchanted, watching them cross, as if they were floating above the ground, calm and graceful.

We could not understand how two such animals could be in that place, surrounded by villages and people. What struck us was the fact that they looked us straight in the eyes, a truly strange thing for animals, especially greater kudus. At that exact moment, I felt a rush of adrenaline coursing through my veins and an inexplicable but very intense connection with the two animals.

A few seconds later, a truck came hurtling down the opposite lane. At rocket speed, and with no regard for the blind turn it was approaching, it overtook a minibus and headed straight towards us. If we had not stopped at the side of the road to watch the two antelopes, the truck would have surely crashed into us, smashing into our car and killing us.

Those magnificent animals, who had suddenly appeared out of nowhere, saved us.

Luca and I were speechless for a few minutes. Then, a strange awareness arose within me. Those antelopes were more than just animals. They were our guardian angels, and they had intervened to save our lives. I like to think they were the spirits of my father-in-law Guido, who had passed away a few years before, and of all my grandparents who were watching over us.

I had read many stories about the existence of guardian angels and had always been a bit skeptical. Since that day, I am no longer doubtful, and I feel their constant presence in my daily life. Now I know that we are never alone and that the friendly spirits of those who are no longer with us protect us, guide us, and stay by our side.

And being aware of this gives me peace and courage in facing every new adventure in Africa.

The Lioness of Okoikuma

The early days spent in Africa were not easy for me. I was still a city girl, struggling to adjust to the untamed nature of that ancient continent. I missed my family and my Italian life; I often felt alone, and the melancholy for what I had left behind weighed heavily on my heart.

I remember one time when I was in my tent, pacing anxiously around my bed, like a lioness in a cage. We had been at the camp for months, building. Months without seeing or hearing from anyone except Luca and the Maasai who worked for us. I couldn't stand the savannah anymore: its silence, its beauty, its purity. I couldn't stand Kilimanjaro either; I wanted skyscrapers to admire instead!

It seems foolish now, but I was truly homesick for concrete, for city noises, for smog, for the crowds of people I was used to before coming to Africa.

Patrick, our cook, had come to see me in the tent. Finding me in that state, agitated and restless, he called Luca on the radio: *"Iko shida kidogo,"* meaning there's a small problem. *"Come see Antonella!"* I love how Africans prepare you psychologically to face a big problem. When someone says *"Iko shida kidogo,"* you can be sure a disaster has happened!

Luca came to the tent immediately. I was pacing quickly, staring at the ground, circling around the bed. I muttered angrily that I couldn't take the savannah anymore and wanted to leave. After exchanging a few words with me, Luca understood what I needed: he organized for a plane and took me to the chaos and confusion of Nairobi for a few days. There, my heart calmed and my thirst for the crowd was quenched, until I longed once more for the peace of Campi ya Kanzi.

But I didn't always need concrete and the confusion of the First World. I remember another beautiful special day, still vivid in my memory as if it were yesterday. It was July 11, 1997, my 28th birthday, and for the first time, I would spend it in Africa.

Luca told me later that he wanted to celebrate me in an unforgettable way. He wanted to erase every trace of nostalgia or sadness from my heart. He knew how much I cared about celebrating that day surrounded by the people I loved, and didn't want me to feel alone, far from my family and Italy.

So he organized a surprise for me: breakfast in the bush! He chose my favorite place: the rock of *Okoikuma.*

In Maa, *Okoikuma* means turtle. The Maasai had given it this name for its extraordinary resemblance to that prehistoric-looking animal. The rock of *Okoikuma* is granitic in origin. Dark and imposing, it stands proud and silent, like an ancient warrior, in the boundless golden plains.

I discovered that it dates back more than 20 million years. The thought that it had always been there, since the beginning of time; the thought of how many wild animals had visited it over the millennia, how many eagles had soared over it, how many lions had taken shelter there, always impressed me and always stirred in me a sort of reverential fear whenever I prepared to climb it.

The first time Luca took me there, I remember thinking that place had something sacred and spiritual about it.

It could be reached via a dirt road that crossed lava flows, forests, and plains, dotted with wild animals, acacias, and baobabs. We left the Land Rover at the foot of the rock and in 10 minutes, following a path traced by animals, we reached the top.

It was beautiful to get there on foot; we often walked on the tracks of a leopard. From the top of *Okoikuma*, it was possible to enjoy the satisfaction and excitement of conquering something primordial and wild. And every time we climbed it, we felt the same intense emotions. It was as if it were always the first time.

Nothing prepares you for the splendid panoramas enjoyed from

Okoikuma. The sensation from the top, overhanging the golden plains, if we closed our eyes, was like flying. The wind blew strong, constantly, ruffling our hair. We got lost in the immensity all around and Kilimanjaro stood still and vast before us, the usual reassuring presence of our African days. A sense of vertigo mixed with euphoric excitement took hold of us; we were silently absorbed in contemplation of the wonderful land of Africa surrounding us.

What struck me most was the extreme variety of colors and of the plains below us. From the deep green of the forest to the light blue of the river; from the ochre, yellow of the savannah to the brick red of the earth. Sometimes, with binoculars, you could see elephants drinking from the river in the distance: tiny black dots breaking the monotony of the plains.

The splendid African sky served as a backdrop. Intense blue and all encompassing; its heft could be felt overhead and seemed to press you against the earth. The infinite African sky boasted floating clouds as far as the eye could see – white with flat bases, soft and fluffy, solitary and calm like *dhows*, the typical wooden boats with triangular sails used by African fishermen in the Indian Ocean.

From the top of *Okoikuma*, you felt as though you could see into the infinite horizon. From there, we seemed to dominate the world. We felt the untamed spirit of the lion enter us, and we felt invincible.

One day Luca said: "If there is a heaven, it must be like *Okoikuma*!" And I agreed with him immediately.

The morning of my birthday, my husband exclaimed: "Get in the car, we're going on safari!" Meanwhile, he had arranged for two of the Maasai who worked with us, Joshua and Tuntan, with another Land Rover, to go to *Okoikuma* and set up a breakfast table before our arrival.

Of course, I had no idea what he'd planned. It would have been a beautiful surprise. Luca had planned to take me to the top of the rock where we would find the table set for us. Tea and coffee, eggs and bacon cooked on the spot over an open fire, fresh fruit, yogurt, and cereal and a delicious chocolate cake with candles. Naturally, there would be a bottle of chilled champagne! What better way to celebrate my birthday?

We were almost at *Okoikuma* when Luca received a radio call from Joshua. He said: *"Iko simba na watoto kwa mulima. Simba kali! Tutafanya nini?* – there is a lioness with cubs on the rock. She is very angry! What do we do?" A lioness and her cubs had decided to rest on the rock of *Okoikuma* that very morning. From there, she could dominate the plains below. Perhaps, Mama Lioness wanted to see if there were gazelles nearby to feed her and her cubs. After all, it was breakfast time; the lioness and her cubs were probably hungry too!

Luca told Joshua and Tuntan to leave immediately and not disturb the lioness. The Maasai replied that they had already set up the table, but the lioness continued to roar nearby. What should they do? Luca told them to leave everything there; they weren't to make the lioness angry! He would go later to retrieve the table.

As the Maasai descended from *Okoikuma*, the lioness began to roar more and more loudly and leap towards them. Joshua and Tuntan then started running until they reached the Land Rover, where, finally, exhausted, they found refuge.

So, no breakfast on *Okoikuma*! We had to return to the lodge, and Luca was forced to tell me about the surprise breakfast. A lioness had booked the same table, he said, and it was therefore impossible to climb up top and have breakfast! Maybe the lioness was offended because she wasn't invited to the feast?

Back at Campi ya Kanzi, everyone gathered around the two poor Maasai, who said that to escape the fierce lioness, they had run for so long that they wore out the soles of their tire sandals!

Still today, the story of the lioness of *Okoikuma* is told by the elders to the Maasai children of the village plains. And still today, Maasai children laugh at the fate of Joshua and Tuntan and how they had to escape from the lioness. What an unforgettable birthday it was for me!

The African Countess

It all began with a phone call from our Italian travel agents: "Two VIP clients are arriving. Please make sure to offer them the best possible hospitality and ensure that their safari at Campi ya Kanzi is perfect and unforgettable!" We had just opened Campi ya Kanzi a few months earlier, and we were, of course, eager to give our guests the safari of their dreams.

So, we made sure everything at the lodge was perfect. The Maasai had worked incredibly hard for two entire days just for the occasion. The Tembo House was in truly wonderful shape. The brass oil lamps were cleaned and polished until they sparkled; the cedar wood floor was waxed; many books about Africa were impeccably arranged in the library; and new ostrich and guinea fowl feathers, as well as sharp porcupine quills, were displayed on the Indian coffee table in front of the imposing stone fireplace. For the occasion, I had created a centerpiece made of pristine ostrich eggs and coconuts, arranged in a flat, round wicker basket from Ethiopia, placed on our wood dining table.

The grass around the camp had been cut, and fresh flowers – callas and white lilies –had been specially flown in from Nairobi with our plane and were placed in the vases of the Tembo House and the guests' tents. Patrick had prepared ginger and coconut pastries and chocolates to offer to the guests along with a cup of Kenyan tea as soon as they arrived. The Maasai were wearing their traditional clothes and their finest jewelry and had lined up in front of Tembo House.

Everything was ready to welcome the VIP guests. And finally, they arrived.

Usually, instinctively, the Maasai can tell within five minutes whether the arriving guests will be pleasant and easy-going or demanding and difficult. It's an art they have acquired with time and experience and is certainly also based on their innate instinctiveness and intuition. I believe that with the African Countess, they understood in less than one minute who they were dealing with. I'm sure they realized immediately that these were not kind people, let alone pleasant or likable.

It was a 40-year-old couple. Elegant and tan, they had an arrogant and disdainful attitude. Unlike the guests who usually arrived at Campi ya Kanzi, when we went to meet them, they didn't smile. They were Italians who had just stayed in Malindi with friends who had a villa there. They had flown directly from the coast to our landing strip in the savannah on a private plane.

When they landed, they were still wearing swimsuits and a *kikoi* (a traditional cotton wrap of bright colors typical of Kenya). I remember thinking that their attire did not seem appropriate for a safari, but, with the travel agent's recommendations in mind, I approached the car, pretending not to notice how they were dressed, and welcomed them with a smile.

Even today, I cannot understand why these people had decided to come on a safari in the heart of wild Africa.

The safari was a disaster! But it wasn't due to any fault of ours or the inefficiency of the Maasai; we all went above and beyond to make that safari unforgettable for our guests. But those people loved nothing about Africa or Africans.

The first words the lady politely said to me were that she was a countess from Rome used to a different lifestyle, and that she couldn't stand flies, dust, insects, and Africans. I remember being struck by the fact that she had lumped animals and African people into the same category. For her, there was no difference.

I was absolutely disgusted by the racist and superior attitude these people had towards the Maasai and felt embarrassed for her. I immediately went to the kitchen to apologize to our staff for the way those people were treating them.

Katito, a wise Maasai woman, told me: "You know, we saw right away that these people's hearts are as cold and hard as a stone!" Then Pashiet, a proud tracker, chimed in. "Don't worry. This is just work for us. We'll treat those people like eggs: we'll move them from place to place with delicacy and care, being sure not to break them so as not to dirty our hands!" Pashiet was absolutely brilliant! With a single phrase, he managed to lighten the mood, and we all began laughing together in the kitchen instead of feeling angry and offended. He had lifted everyone's spirits, and it would now be much easier to deal with these people.

During her stay at Campi ya Kanzi, the countess never went on a safari. She said she would get dirty and that her makeup would be ruined. Instead, she loved to spend hours sunbathing on the terrace of the Tembo House in her bikini.

By the end of her stay, she was very tan. So tan that Katito, naively, came to me and asked: "Have you seen how dark that lady's skin is? It's just like ours... Could she possibly have African relatives? Maybe her mother was born in Africa?"

It was unbelievable! The countess felt so distant from Africa and Africans. If only she had imagined that Katito and the others thought that she might even have an African mother...

I don't know why I did it, but when the guests left, I told everyone that the countess had confirmed to me that her mother was born in Africa. The Maasai didn't seem very impressed, but I considered it a sort of personal revenge against the countess's arrogance and disdain towards the Maasai.

A Slightly Unusual Wedding

After completing the construction of Campi ya Kanzi and finding in the Maasai the warmth of a family, I decided to marry Luca to honor our relationship and love surrounded by our family and friends.

Naturally, we had to marry in Kenya, in our new home, among our new people.

Convincing our respective Italian families (especially mine!) to come to Africa for the wedding was no easy task. In the end, over 30 people flew from Italy to Nairobi and another 60 from all over Kenya at the beginning of December 1999 to participate in the grand event: the wedding of Antonella and Luca, the "white Maasai!"

I planned everything down to the smallest detail, and it took me months.

First, Luca and I chose the place. It had to bring joy to the soul and have a boundless view of the African plains and the immense Kilimanjaro. We wanted a natural altar and a canopy of leaves and flowers over our heads.

After exhaustive searches, we chose a spectacular hill that the Maasai call "the Soft Hill" because it is covered with green grass so smooth it seems like velvet. At its summit stands a tree with red flowers, an *erythrina abyssinica,* cherished by the Maasai, who make good luck amulets with the seeds of its fruits.

It is a tree with grey, dry, and wrinkled bark like the skin of an old, wise elephant, with twisted branches stretching toward the sky like fingers of hands reaching into the heavens. It only blooms once a year, but it is impossible to predict exactly when; it has splendid, unmistakably red flowers – the same red as the *shukas* of the Maasai warriors.

The Chyulu Hills are full of these trees, and when they are in bloom, from the plains below, thousands of red dots can be seen as a contrast to the vibrant green of the grass and the deep blue of the African sky. It is a breathtaking spectacle of beauty. In my darkest days, when I feel sad and empty, I close my eyes and search through my memory for the vision of the blooming *erythrinas* in the Chyulus; then my heart opens, breathes, and beats again with renewed strength and energy.

Under that splendid tree, the Maasai had built a natural altar for us out of stone, grass, and flowers. Next to it stood a cross made of wild olive, tied in the center with moss and lichen found in the forest and decorated with white orchids. We had prepared wooden benches for the guests and placed bales of straw covered with flame-red Maasai *shukas* for the bride and groom.

My wedding dress was made by a friend of mine, an Irish designer, who had lived in Kenya for several years. We designed it together. It was white, with a long, simple skirt and a small train. The corset was embroidered with ostrich eggs, Maasai beads, and dried plant seeds found on the hills, symbols of good luck and fertility. On my head rested a garland woven with wildflowers and orchids picked in the forest. The bouquet had white calla lilies and long white ostrich feathers tied with a suede ribbon embroidered with Maasai beads.

That year, the rains had lasted longer than usual. Sometimes, in those days, the sky would darken and begin to rumble; then heavy clouds would unload on the Chyulu Hills, making the ground muddy and slippery. I was very worried it might rain on our wedding day. Ninety guests on a hill, with no shelter under the torrential rain that I had only seen in Africa, would not have been a pleasant sight. It would have been a disaster! I had to do something to make sure that didn't happen.

The day before the wedding, without anyone knowing, I called the *laibon*, the Maasai village's witch doctor, and begged him to help me. "*Mzee* [wise old man, a highly respectful term in Kiswahili] I need your help. I know you have many magical powers. It cannot rain on my wedding day tomorrow! What should I do?" The old *laibon* was hunched over his ebony stick, dressed only in a splendid red and blue *shuka* and a hand-painted goatskin cloak. He was adorned with a multitude of necklaces and earrings made of colorful beads and dry seeds. He looked at me for a long time with deep, lively eyes, which were strangely young for his age. It seemed to me that he was scrutinizing my soul, searching for a sign confirming my sincerity and my blind trust in him and his powers.

Then he clasped one of the many jewels around his neck in his thin, calloused hands, an amulet made of dried red *erythrina* seeds, and, revealing nearly toothless gums, he said, "*Naseryan* [the name the Maasai had given me] take this necklace and bury it in the exact place where you will get married, tonight. Tonight, I will make a sacrifice for you to *Enkai*. I will slaughter a black goat and ask him to help us. Tomorrow, on the day of your *arusi* (wedding), it will not rain. I promise." He walked away, with dignity and sacredness, slowly, leaning on his stick, his beautiful shaven head gleaming in the golden light of the sunset, leaving me doubtful but fascinated by his words.

I did as he said, knowing I was doing something extremely irrational and unacceptable by the standards of the culture I came from. But wasn't I in Africa? Here, western rules did not apply; here, people believed in nature and witch doctors, so it was definitely worth a try!

The next morning, December 11, the day of our wedding, the sun was shining, and snow-capped Kilimanjaro stood tall in all its majesty against the blue sky. Not a single cloud smudged the beauty of that sky. We were fortunate and blessed with such a sight, and I could not help but wonder whether the *laibon* had really helped us!

The velvet green grass of the hills was speckled with pink wild cyclamens, unrolling like a precious carpet under our feet. The *eryth-*

rina tree on our hill was miraculously in bloom: the only one of all the trees on the hills. *Enkai* was with us.

In the morning, more than 60 guests had arrived from all over Kenya with their small planes and landed on our runway, to the amazement and excitement of the Maasai, who saw the sky dotted with *ndege* (birds or airplanes in Swahili). Some of them never saw an airplane before; they were both intimidated and fascinated by what they saw.

Since it was impossible for me to accommodate everyone at Campi ya Kanzi, I had organized a temporary camp behind our lodge. Two days before the wedding, about 30 tents had been pitched, bucket showers hung from acacias, and latrines had been dug for the so-called "pit toilets."

In addition to guests from Kenya and Italy, the Maasai had to be in attendance. They had been preparing for this event for months. They arrived on foot, some even from Tanzania, dressed in their splendid traditional clothes, to witness the wedding of Luca and Antonella.

We had decided to marry at sunset, when the sky turns red, and Kilimanjaro appears even more beautiful and imposing.

Before the ceremony began, about 200 *morans*, the Maasai warriors, arrived. They came dancing and singing their festive songs. They jumped high, in their unmistakable manner, gracefully and proudly waving their long ochre-dyed braided hair in the air. The bright red of their *shukas* and the vibrant colors of their jewelry contrasted with the blue of the sky: they were a unique and stunningly beautiful sight. I will keep that spectacle in my memory, as something rare and precious, to be treasured forever.

The women also dressed festively with their large round necklaces around their long, slender and elegant necks, their shaved heads gleaming in the sun. And there were the children, many children, who sang happily, took our hands, and laughed with us.

Since the place where the wedding would be celebrated was about half an hour's drive from the lodge, I decided to dress at the last moment, at the foot of the hill. A mirror had been brought under an acacia, and surrounded by cows and goats, in the company of two Maasai shepherds and a *moran* who watched me attentively and curiously, my hair rumpled by the wind, excited and happy, I put on my wedding dress.

I then walked up the Velvet Hill to meet my groom and unite with him forever. I chose to be accompanied to the altar by the Maasai women; there were eight of them, and they crowded around me, as if to protect and care for me. They sang as we climbed the hill. Two of their daughters, two beautiful little girls, were my bridesmaids and held the train of my dress. Excited, silent, and proud, with my bouquet in their hands, they filled me with immense tenderness.

The ceremony was very moving; it was conducted in Italian, English, and Kiswahili so that everyone could participate. After we said our "I do's," before descending the hill, I spotted an old Maasai, a few meters away from the tree with the red flowers, standing still and watching us; he was dressed in cowhide and adorned with ancient jewelry. He leaned on his ebony stick and silently observed what was happening. I felt his heartbeat with mine, and for a moment, it was only me and him, in the enchantment of that precious moment. His anklets tinkled in the breeze at dusk as he walked, looking at me intensely, and left, disappearing forever into the mystery and magic of that place and that moment. But that memory would forever remain in me.

At the end of the ceremony, we all returned to the lodge. The sun had set, and a celebration awaited us – Italian, Kenyan, and Maasai friends gathered around bonfires, candles, and oil lamps, with music, singing, and dancing beneath the endless African starlit sky.

The next day brought the most beautiful surprise: the Maasai elders came to our camp to marry us with their own ritual. They had sewn traditional wedding garments for us, dressed us in them, and conducted their marriage ceremony. They brought us ritual gifts. For me, a *kibuyo,* a dried calabash decorated with Maasai beads, symbolizing fertility, and for Luca, a *rungu*, an ebony club symbolizing the authority and power of the head of the family.

They made me promise to always be faithful to my husband, to obey him, to build a beautiful home for him, not to lose any of his cows or goats, and to fetch water and wood every day. They told Luca to always protect me and stand by my side, and to give me many children! It was deeply touching and affirmed that we had been accepted by the Maasai and were part of their splendid family. However, I remained doubtful about my ability to keep my promise of obedience to Luca...

When the guests had departed in their planes and cars, tranquility returned to the velvet hills. But something had changed: we felt truly at home among genuine friends, and the realization grew within us that we would never be alone again.

We were a family now: Luca, the Maasai, and I, embraced by their warmth and adopted by them.

Months later, Luca told me that, on the velvet hill, under the red-flowered tree next to our stone altar, the Maasai often spotted a pair of lions – a male with a grand mane and a beautiful female – resting. They came especially at sunset. I like to believe they climbed there to enjoy the magnificent view of their Africa, drawn by the same magic and good energy that we experienced deeply on our wedding day. I like to think of them as our friends. Knowing of their presence at our wedding site makes it even more special and dear to my heart.

A Stroll in Paradise

Two years had passed since our wedding day in Africa, years that had flown by as they do when life is filled with joy and ease. Those years were exhilarating and intense, characterized by adventure and magic, joy and the occasional bout with sorrow. With each passing day, we felt more at home, and the pangs of nostalgia for my Italy became increasingly rare.

In October 2001, I became pregnant. I had a profound longing for this child, while Luca was more apprehensive. Our lives were about to change forever, and I was excited and overwhelmingly happy about it.

It was the beginning of November, and I was one month pregnant with my daughter Lucrezia, though I didn't yet know it. But I suspected it. It was a strange, profound feeling, an intuition that only a mother can have. I felt strong and invincible, full of energy and optimism, and at the same time, fragile and in need of protection, without truly understanding why.

Strangely, I felt the urge to take a walk on the hills.

During our years in Africa, Luca often teased me about my laziness. It was my nature; ever since I was a child, I disdained intense physical activity, especially when it involved hiking up hills. I had always preferred to spend my time reading poetry, writing, playing the guitar, listening to music, meditating, or practicing yoga – activities for the mind rather than the body.

This time, however, was different. I had a great urge to climb the hills; I wanted to feel the wind in my hair and the warmth of the sun on my skin. I felt the need to merge with the beauty of the universe, to connect with its magical energy.

I needed to breathe in the scent of the wildflowers and hold ferns and orchids in my hands. I wanted to rest and observe in respectful silence the sunbirds as they eagerly drank the nectar of the *leonotis*, those splendid orange flowers that grew along the dirt road leading us into the heart of the Chyulu Hills. What delightful creatures those little birds were! The colors of their feathers were vivid and dazzling – crimson red, purple, golden yellow, emerald, green, indigo blue, and velvety black – shimmering in the sunlight. They were small, almost tiny, graceful, and gentle; they fluttered delicately from flower to flower, eagerly sipping the sweet nectar with their long, slender beaks. It is no surprise that their name links them to the beauty and warmth of the sun.

Our hike was along a new path that led into the heart of the forest, up one of the highest peaks of the Chyulu Hills. The Maasai and Luca had just discovered it. It led to such a splendid place, that Matasha had nicknamed it "Peponi Walk," a walk in Paradise. Luca had found it by following the trail of an elephant he had seen sliding down a hill on its backside, as if enjoying "skiing" on the grass.

That day, when Luca suggested we go on a walk in paradise, I did not object and accepted with unusual enthusiasm. After six months of his insistence, he had finally convinced me, and without much effort this time.

That afternoon, we took the Land Rover, and along with our dogs, we headed up the hill to discover paradise. The dusty road leading to the edge of the forest was spectacular. The never-ending views took our breath away.

To the south, below us, the gentle hills of Tsavo and Taita slowly unfurled, their characteristic bronze and burnt-earth colors reflecting the dust of Africa. To the west, the pale pink plains of Amboseli gave us a real sense of the vast African expanses, and right in front of us stood Mount Kilimanjaro, immobile and imposing, with its snowy peaks like whipped cream atop a grand birthday cake.

The road continued, winding and climbing between green grass and colorful mountain flowers, surrounded by trees with twisted, rugged trunks, like solitary modern art sculptures, and precious ebony bushes. At one point, the path joined the old road built by the British in the 1950s. It was meant to connect Tsavo National Park to the Chyulu Hills, but it hadn't been used in many years. Those pioneers had done an excellent job, creating a trail that ran through the middle of the hills, splitting them in two, allowing one to fully enjoy their beauty from north to south. From up there, we were treated to the incredible spectacle of the Chyulu Hills, Hemingway's Green Hills of Africa, soft and gentle, perfectly rounded like a woman's full breasts.

As we started along the Englishmen's Road (as we called it), the ferns timidly made their appearance alongside the fabulous *erythrina abyssinica* trees, characterized by the beauty and intensity of their crimson red flowers.

We often stopped to watch the little birds flitting from flower to flower along the roadside, or to gaze at solitary eagles soaring proudly in the endless African sky. Up there, we often encountered giraffes, eland, and kongoni. If you looked carefully through binoculars, you might spot the shy and graceful mountain reedbucks, agile little antelopes, always astonishing us with how quickly they could scurry to the tops of the hills. Sometimes, at dusk, we would come across a solitary buffalo or an elephant stubbornly making its way to the top of the Chyulu Hills to drink from the watering holes that formed beneath the towering forest trees.

That day, we left the car at the edge of the forest, with our backpacks on, accompanied by our happy, tail-wagging dogs, and ventured into paradise. Matasha, our most faithful and experienced Maasai tracker, led the way. Precisely and with no hesitation, he cut through the tangled vines and shrubs with his *panga*, a long, sharp machete, allowing us to weave our way through the impenetrable forest.

The first part of the walk was very steep and tiring. For a good half hour, we climbed towards the summit of one of the highest hills; there was no path to follow except the narrow and winding one created by buffalo and other wild animals. We struggled to climb under the scorching hot sun.

Matasha and Luca often stopped to wait for me, patiently, teasing me good-naturedly about how slow I was. They pointed out the most beautiful flowers and plants, as well as the tracks and droppings of some animal that had passed through shortly before us. This was their way of distracting me from thinking about how tired I already was. When I complained, Luca would respond, "You'll see, when you get up there, it will have been worth it."

I decided to trust him; I felt that the place he would show me would be incredibly special. Finally, after a great deal of effort, we entered the thicket. Now we walked on level ground, sheltered from the sun, and it was much more enjoyable. After about an hour of walking, we reached the highest peak of the Chyulu Hills, over 2,000 meters in altitude.

The forest up there was wonderful and magical: enormous strangler figs, vines, moss, lichens, orchids, ferns, and wild jasmine adorned it. It was strange how it really gave everyone the impression of an enchanted forest! The golden light of the afternoon sun filtered through the dense branches. We spotted a turaco with blue and crimson red feathers suddenly flying from one tree to another, like a flash of fire; we stopped, our hearts pounding, hearing the beat of its wings above us.

Sykes' monkeys played, noisily jumping from branch to branch. Their black fur and white collars fascinated me. We walked on and came across the fresh droppings of elephants and buffaloes that had taken the same path that morning. It was thrilling; it was like approaching a mysterious place, discovering a hidden treasure. It was like being a child again, reliving the carefree days of secret adventures!

In the forest we listened, silent and tense, to the countless sounds surrounding us, and our eyes never tired of looking around. We had a strange feeling. That place inspired infinite respect and a sense of sacredness that left us speechless.

After about an hour of walking, we finally arrived at the place Matasha called "Peponi." It was a small clearing where a huge strangler fig stood majestically. From its branches and roots hung a myriad of vines covered with moss and lichens of such an intense green that it hurt the eyes. That tree had to be several hundred years old. The sunlight bounced off the emerald leaves and filtered playfully through the branches as the wind gently stirred them: it was a surreal vision. It seemed as if the fingers of those immense branches were calling us, cajoling us to come closer.

That enormous tangle of shoots and vines created a sort of structure, a kind of cathedral in the Forest. About 30 meters high, the vines and twisted branches formed the columns and architraves of our sacred temple; the smaller, crisscrossing trunks, moss, and lichens were the precious capitals and sculptures that adorned the columns. The tangled leaves, in all their bright shades of green, punctuated by the white of the orchids and wild jasmine, were the dome and the carpets of our Cathedral, made even more magnificent by a sort of natural fresco.

Hypnotized, we advanced on tiptoe so as not to break the spell of the moment, towards those enormous figs, to finally touch them, stroke them, and feel the moisture of their trunks, the roughness of the bark, their scent, and their ancient breath.

I closed my eyes and lingered there, in silence. A deep urge to meditate washed over me, to bask in the peace and serenity that had quietly settled into my heart. I felt an overwhelming sense of well-being, detached from everything and everyone. I absorbed the magical energy of that place, listened to my soul, and reveled in its joy. I knew these were rare and precious moments, and I hoped they would last as long as possible.

Luca loved coming up here. He often brought books of poetry, sitting beneath the fig tree for long stretches, reading and letting the voice of his thoughts drift away in the wind over the Chyulu Hills.

This was a sacred place for the Maasai, where they performed sacrifices during droughts. They would bring a black goat under that very tree, slit its throat, and pray to their God, *Enkai*, for rain. Since we had arrived, the Maasai elders had performed the rain sacrifice only twice, and incredibly, it had rained the day after each time.

It was getting late, and we needed to return to the camp before nightfall. As the sun set, we walked back to the car in silence, each of us lost in our own thoughts, with the awareness that something within us had shifted. After all, we had been to paradise, an experience that not everyone is fortunate enough to have, and one that would linger in our souls forever.

Upon returning to the camp, I realized I had lost my sunglasses. They meant a lot to me, and living so far from Nairobi, getting another pair wouldn't be easy. The next day, I asked Matasha to look for them. He returned in the evening, tired and disappointed; he had retraced the same path we had taken together, back and forth. Unfortunately, he did not find anything.

That night, I had a strange dream: I was in the forest, just as I had been the day before, with Luca and Matasha. I was walking among the enormous trees, in silence. Suddenly, I stopped under a fig tree larger than the others, its roots covered in moss and delicate wild jasmine flowers. I felt an inexplicable urge to search for something in that spot, and to my surprise, I found my sunglasses! I woke up with a strange feeling of euphoria and disbelief.

In the days that followed, I could not stop thinking about the dream. A few days later, I felt compelled to return to the forest, to Peponi. I retraced the path of my dream –the same walk, the same trees, the same flowers. When I reached the great fig tree from my dream, I stopped. I looked down at the ground, among the moss and jasmine, and there they were! The sunglasses were exactly where they were in my dream!

I cannot explain what happened rationally, but in Africa, I have learned to let go of reason and trust my instincts and intuition.

Luca often tells me that he finds buffaloes and elephants there, beneath the cathedral. One time, a buffalo stopped just a few meters away from Luca, looking at him with curiosity. It approached as if trying to understand who had come to this paradise; then, it vanished with a gentle puff and a peaceful retreat, without the usual aggression typical of buffaloes.

Peponi is indeed the most fitting name for the place in that enchanted forest where Luca and Matasha took me!

The Crater of Wonders

The Ngorongoro Crater in Tanzania had always been one of the places I longed to see since my arrival in Africa. Its reputation for being breathtaking – a true heaven on Earth – preceded it. Simply put, it is considered one of the Seven Natural Wonders of the World, which speaks to the beauty and magic of that place.

The opportunity finally arose when two of our Italian friends, avid Africa aficionados who had visited Campi ya Kanzi several times, asked us to accompany them on a visit. It was April 2002. The season was ideal because the rains were beginning, and we were certain to find the crater teeming with wildlife. Moreover, the landscape would undoubtedly be covered with green grass and colorful flowers, making for a true spectacle of nature. Visiting the Ngorongoro Crater in April, during the low season, also meant avoiding the hordes of crowded minivans typically filled with tourists, so this made the idea all the more appealing.

The only minor issue was that I was seven months pregnant, so I would have to bring my big belly along on safari! This did not scare me much; the desire to see that place was so strong that I mustered the courage and decided to go on the trip. Was it reckless? Perhaps, but my bravery was rewarded as everything went smoothly, and I saw beautiful things, making it an unforgettable experience.

We left in our six-seater plane on a sunny morning. Given the amount of luggage, the number of passengers, and the pilot, we decided to make two trips. After a half-hour flight, we landed in Taveta, a small village on the Kenya-Tanzania border. The boys decided to go ahead on the first flight. When my girlfriend and I arrived in Taveta, a car was waiting to drive us to Moshi, from where we would fly to Ngorongoro. Our husbands were supposed to be waiting for us in the Land Rover. However, as soon as we landed, I spotted someone at the edge of the airstrip, but Luca was nowhere to be seen.

Tired of waiting for us girls, he had characteristically decided to take a walk to stretch his legs. I hoped he would be waiting for me at the Kenyan customs on the Tanzanian border, but when we arrived, there was no sign of Luca. Unfortunately, he had kept walking without realizing he had already crossed into Tanzania. The problem was that I had his passport, which had to be stamped even though he wasn't there, and this was easier said than done.

The customs building in Taveta was decrepit, dirty, and sad. But inside, the African officials were in full uniform, smiling, and seemingly very efficient. I gathered my courage, entered with my passport and my husband's, and began to enact one of my favorite roles, naturally performed in Kiswahili: the poor pregnant wife with a reckless husband who had wandered off, asking for the help from the wise *wazee* (elders, worthy of respect in Kiswahili) to save her from a really difficult situation. What can I say? Africans love a good bit of drama: after just a few minutes, they were laughing heartily, shook my hand, and stamped both passports without asking any questions! This was exactly one of the reasons I loved Africa so much. And where else but in Africa could something like this happen?

Fortunately, Luca was waiting for us just a little further ahead, blissfully unaware of what had just happened. After we got him into the Land Rover, we continued our journey for about an hour until we reached Moshi. From there, we took another small six-seater plane to the top of the Ngorongoro Crater. As soon as we landed, the crisp, refreshing mountain air made us shiver. We were over 2,000 meters above sea level, surrounded by clouds. The view of the crater more than 600 meters below was magnificent. We couldn't wait to explore it the following morning.

There are several lodges on the rim of the Ngorongoro Crater, and others just outside of it. The most beautiful to me is undoubtedly the Ngorongoro Crater Lodge, where we decided to stay. When we arrived, we were greeted by a very professional and friendly South African manager who offered us a welcome cocktail. After explaining the various things we could do over the next few days, he showed us to our bungalows.

Designed by a famous South African architect, the lodge is a blend of French Baroque and African décor. Rich golden frames holding ancient, valuable paintings adorned the walls, and imposing Rococo mirrors stood beside intricately carved African masks, often depicting intense and curious facial expressions, reminding us of fierce battles and sacrifices to the gods.

Entire walls of wood paneling created a pleasant feeling of warmth and welcome, alternating with enormous windows that opened fully to the infinite beauty of Africa. Fine brocade fabrics covered the sofas, complemented by precious velvet drapes on the large windows. Rare pieces of African art sat proudly on antique French crystal tables, and soft zebra skins lay beside splendid Persian rugs on the mahogany wood floor. Everything was exquisitely paired in colors and shapes, making the lodge elegant, harmonious, and comfortable and truly one of a kind!

The lodge consisted of a central communal area for dining and relaxing after safaris, along with 15 bungalows for guests. The structures were built of masonry and African cedar wood. The roofs were made of woven banana leaves, creating a stunning effect for those who came to visit. I immediately fell in love with the place. The fact that such a rich and luxurious construction had been erected in such a wild and primordial landscape created a contrast that deeply fascinated and moved me.

Our accommodations were superb. In the bedroom stood a large four-poster bed with a precious damask bedspread. Large windows offered a magnificent view of Africa. The bathroom was spacious and comfortable, with painstaking attention to detail: huge vases of fresh roses delivered daily, scented soaps and excellent body creams, and linens of the finest quality. In the center, the cast-iron bathtub with wrought-iron feet offered a view of the infinite beauty of the Ngorongoro Crater. A fireplace beside the bed, where a crackling fire was lit every evening at sunset by our personal attendant, completed the picture, making it all truly romantic and evocative. That evening, after a delicious candlelit dinner, we retired to our rooms.

The next morning at dawn, we drove down to the crater floor on a muddy dirt road. We were finally on safari: the spectacle that unfolded before our eyes that morning was of such beauty that it remains vivid in my memory. The crater's slopes were dotted with countless colorful flowers gently swaying in the breeze, and a precious layer of tender, vibrant green grass covered the ramps like a carpet.

Colorful butterflies and charming little birds fluttered from flower to flower. In the middle stood a sparkling soda lake, Lake Makat, filled with pink flamingos, which occasionally took flight, allowing us to tap into a whole new range of emotions.

It is said that around 25,000 animals live permanently in the crater. Considering its small size, about 260 square kilometers, you can imagine the incredible density of animals. The variety is unique: from all the animals of the plains like zebras, antelopes, wildebeest, giraffes, warthogs, and others, to majestic prides of lions with their cubs; from immense elephants to the rare black rhinos; from large buffaloes to hyenas and jackals; not forgetting the timid cheetahs and the magnificent leopards! And then, countless varieties of birds, including pink flamingos, and plants to observe. In short, that morning, in just a few hours, we saw everything one could hope to see on an African safari, in a unique and wonderful landscape.

The only downside to all of this was the number of people. Despite it being the low season, there were quite a few vehicles inside the crater, making it likely to find yourself surrounded by 20 minibuses watching a poor lion hunting or a solitary elephant bathing.

Being accustomed to the space and exclusivity of our beloved Chyulu Hills, this was hard for us to accept. I remember we had stopped for a picnic at lunchtime, and suddenly, at least 10 other vehicles arrived at the very same spot. This was the only picnic site in the park, so everyone had to eat their lunch there, like it or not, surrounded by so many others! This put a bit of a damper on our experience at the Ngorongoro Crater. We were aware of being spoiled, as we were used to having over 1,000 square kilometers at our disposal, with no one else to disturb our wildlife sightings. It was difficult to accept a crowded situation like that of the crater.

At sunset, upon our return to the lodge, a surprise awaited us: in our bungalow, the fire was lit, and on a table in front of it, there were two glasses of fine port wine and some Swiss chocolates. A path of rose petals led from the door to the bathtub, filled with hot water and scented suds. What could be better than this? I immediately immersed myself in the tub to contemplate the fiery sunset and the vastness of the crater below us; to this day, that was the best bath I've ever had in my life! That evening, the dinner exceeded our greatest expectations.

The next day, it was already time to return home to Campi ya Kanzi. Unfortunately, the weather was not promising. It had started raining heavily at dawn, and our pilot could not come to pick us up at the Ngorongoro airstrip. So, we drove on a red dirt road that was now flooded, from the lodge to Manyara. From there, we were supposed to take a small plane and land at Kilimanjaro Airstrip, near Arusha.

During our journey in the Land Rover, the driver, utterly incapable of handling the vehicle, managed to skid into a ditch beside the dirt road, leaving us suddenly teetering on one side in the car. The only way out was through the rear door. It wasn't easy for me: with my belly, I wasn't particularly agile, and what was a simple movement for the others was much more challenging for me. It took me a while, but after a few minutes, I managed to get out of the car. Standing in the pouring rain, I waited as the others pulled the vehicle out of the hole. It was quite frightening. For a moment, I regretted coming all this way!

Finally, after countless obstacles, we reached our destination.

Fortunately, upon arriving in Manyara, the weather improved, and we were able to take off. After about an hour of flying, we landed safely at Kilimanjaro Airport. From there, we caught a commercial flight to Nairobi, arriving a couple of hours later. It had been a long day, and we needed to rest.

That night, we stayed with friends in Nairobi, and the next day, we flew back to camp, still tired but happy from the beautiful adventure we had just experienced.

To this day, Ngorongoro remains the crater of wonders for me, despite the challenges we faced during the journey. I hope to return one day, with my children. I know they would love to see that abundance of animals and the marvelous nature that is their home. And, after returning to the Chyulu Hills, they would have a greater appreciation for the uniqueness and purity of the place they call home, feeling fortunate and privileged to enjoy one of the last remaining corners of true, wild Africa.

Lucrezia, Macho Maridadi

At last, after a serene, uncomplicated pregnancy, that day arrived, and it was one of the most beautiful of our lives. My daughter Lucrezia was born on a radiant July morning at the Nairobi Hospital. It was 10:05 a.m. on July 12, 2002. My birthday falls on the 11th of the same month, and I couldn't help but see the fact that my daughter was born just one day after me as a sign of good fortune.

Of course, her birth could only be in Nairobi, for she is a child of Africa, and her father and I are both proud and honored that she is. The doctors and nurses were kind and highly professional. Lucrezia was a beautiful baby: at birth, she weighed nearly four kilos and was 51 centimeters long. Even in the delivery room, what struck us most were her splendid, deep blue eyes.

When Luca placed her on my chest, she looked at us with her wide, velvety eyes, studying our souls with an intensity that seemed to give her the essence she needed to recognize her mother and father. This phenomenon, known as *imprinting*, happens with animals as well; it is the most crucial moment, the genesis of everything. It is when the baby and mother recognize each other, breathe in each other's essence, and, I like to believe, gaze into each other's souls. This moment forges an unbreakable bond between mother and newborn, a bond that remains unchanged no matter what. It is intimate and special, unique and extraordinary.

The same was true for Lucrezia and us. What her father and I felt at our first encounter with her is both unrepeatable and unforgettable.

Even now, her beautiful grey-blue eyes captivate all who meet her. She truly deserves the Maasai name she's been given: *Macho Maridadi*, Beautiful Eyes.

At just five days old, Lucrezia, Luca, and I flew from Nairobi to Campi ya Kanzi to begin our life as a family. The Maasai were eager and joyful, waiting at the airstrip to welcome their new friend with open arms. As soon as we landed, a curious, graceful giraffe approached the Land Rover to observe my daughter. I like to think I saw a shadow of affection and emotion in its gaze, framed by long, dark lashes; it looked upon Lucrezia with tenderness.

Lucrezia's first home was a tent, but this did not seem to trouble her at all. At that time, Luca was still building our house, so we spent our early days in a tent. Though it was spacious and filled with comforts, it was still a tent! Today, we live in a beautiful home, but Lucrezia often returns to her tent, accompanied by her father. She says it brings her more peace and that nature speaks to her there.

How fortunate my daughter was! Each morning, she awoke to the sight of the snow-capped Kilimanjaro. Though she was so small, she was already a friend to elephants, zebras, gazelles, giraffes, buffaloes, lions, and hyenas. Her blue eyes were filled with the boundless African skies; the warm desert wind tousled her soft, blonde hair, and her fragrant skin gleamed with shades of bronze and the softness of silk. She breathed in the exotic scents of spices and clay soil, and her hands wove golden savannah grass and forest orchids into garlands for her head. The lion's indomitable spirit lived within her, and her free soul soared like the eagles high on the hill. Freedom and nature were her playmates, and at sunset, after an exhilarating day, she returned home to surrender to the cool embrace of another starry African night. Mother Africa cradled her with love and sweetness until she fell into a peaceful slumber.

While pregnant with Lucrezia, I came across a passage in a book by Karen Blixen that felt as if it was written for my daughter, perfectly capturing my vision of her:

"The children whose parents have loved each other deeply show a fearlessness toward life that those conceived in cold beds cannot know. They are like cherubs on ancient bas-reliefs riding on lions, spurring the mighty lord of the desert with their pink heels, and pulling at his dark mane. The lion has been their guardian and their friend, and when they meet him again in life, they recognize their old playmate and laugh."

This is how I think of my daughter: guarded by lions and cradled by giraffes. Fearless and radiant, courageous and joyful. The buffaloes watched over her, and the hyenas sang lullabies to soothe her to sleep, making her smile with contentment. As she grew, the elephants were her silent and solemn sentinels on solitary walks across the boundless plains, standing watch for as long as she desired.

On Lucrezia's first night in her tent, as I nursed her, the lions roared just a few meters away. She suckled peacefully at my breast, serene and unfazed. I like to think they had come to welcome her to this magical land of Africa.

From the beginning, Katito, a Maasai woman my age with four children of her own, helped me care for Lucrezia. My daughter became her fifth child. Katito had worked with us for some time, and she was an excellent cook. She had an affinity for making risotto alla Milanese, and her talent in the kitchen was unparalleled.

Katito's first child, Layoni, was born when she was just 18, which was not uncommon among the Maasai. Girls marry very young, often to men who are much older, chosen by their families. These men, who are held in high regard in their communities, take innocent and inexperienced girls as wives, often leaving little room for romance and love.

Katito's mother was Kikuyu, but her father was Maasai. She confided in me that she had always felt entirely Maasai. This courageous woman was dignified in accepting the man chosen for her – Sonkoi, who was 20 years her senior. Despite his occasional drunkenness and the resulting violence, she persevered with pride and determination, drawing strength from her love for her children.

I had always admired her strength and resilience. Despite her hardships, she possessed a deep sense of humor and a maturity beyond her years. She told the funniest anecdotes from her life with a twinkle in her eye, and she was equally adept at being firm and nurturing with Lucrezia when needed. This is why I chose her to help raise my daughter, and we embraced her as if she had always been a part of our family.

As soon as she saw Lucrezia, she took her in her arms and joyfully exclaimed, "*Jambo na karibu, mtoto wangu!* Welcome to this world, my little one!" From that moment on, Katito and Lucrezia were inseparable. It was love at first sight! Katito sang the first lullabies in Maa to lull Lucrezia to sleep, rocked her when she cried and played with her to bring smiles to her face.

To the Maasai at the camp, Lucrezia was a true blessing, as if she belonged to everyone. This is a testament to the sweet, maternal nature of the Africans: they adore children, who instinctively return their affection.

About 10 days after our arrival at Campi ya Kanzi with Lucrezia, an assembly of elder Maasai women walked all night to meet our firstborn. The next morning, they arrived, bringing gifts and were eager to see, touch, and welcome Lucrezia into their family.

They wanted to present her to the world with a Maasai ritual. Dressed in their finest traditional clothes, adorned with exquisite jewels, they lined up before us. With ancient wisdom in their eyes, they approached Lucrezia one by one, placing a hand on her head, spitting on the ground, and reciting solemn words in Maa. They named her *Namunyak,* meaning "She who will be fortunate because blessed by God." They gifted her a small calabash decorated with beads, bracelets, and necklaces.

Luca and I saw it as a greatly profound moment, full of spirituality and great emotion. At the end of the ritual, the ancient Maasai women turned away, returning in quiet reverence to their distant village. Their sacred task was fulfilled. They could now journey back to their homes in the plains, bringing the news to the other Maasai: they had witnessed the white man's cub and had embraced and welcomed her into their great family.

Katito in Italy

"Kama wewe utaenda kwa Italy, nani atachunga mtoto wetu? When you are in Italy, who will help take care of our little one?" This was how Katito launched into conversation one bright August morning, as she opened the tent where we lived, kissed Lucrezia, and greeted us with her usual warmth.

Lucrezia was almost two months old by then, and it was time for her to meet her Italian family. Grandparents, uncles, aunts, cousins, and friends were all eagerly waiting to see the baby, constantly asking, "When will you bring her to us?"

So, I had discussed it with Luca, and we decided to embark on a journey to Italy.

But now, after reflecting on Katito's words, I began to consider bringing her along, not only to show her Italy and introduce her to our families but also to keep her close to her beloved Lucrezia for the entire month. I asked her if she wanted to come with us, if she could bear being away from Africa for so long. I explained that Italy was very different from her homeland; that there were cars, skyscrapers, smog, and so many people.

Katito had never even been to Nairobi! I was really conflicted about what to do. I feared it might be too overwhelming for her, a negative and unsettling experience. But this brave Maasai woman was enthusiastic about joining us. She kept repeating that she wanted to see Italy, to fly on an airplane, and meet our families. Above all, she was steadfast in her belief: she was convinced that no one in Italy could properly take care of little Lucrezia.

Her enthusiasm and determination were infectious, so I decided to take her with us. Getting her a passport and visa was not easy, but by personally vouching for her, we managed to secure them.

In the days leading up to our departure, Katito was happy and excited. She helped me pack for myself, Luca, and Lucrezia. She didn't forget anything. She was attentive and meticulous in organizing everything we might need for the journey. I did not know what I would have done without her help.

We had to leave early in the morning from the camp to fly in our plane to Nairobi. We would land directly at Jomo Kenyatta International Airport and from there, board a Swiss Air flight to Zurich and then on to Italy.

On the morning of our departure, Katito's husband, Sonkoi, came to our airstrip to bid his wife a final farewell. It was touching. He had walked all night just to see her one last time. It had always seemed to me that there was no tenderness when it came to Maasai couples, but Sonkoi's kind gesture made me rethink that belief. I realized I still had much to learn about this proud culture, and that the relationship between husband and wife was far more complex and profound than I had imagined. Katito wept silently; her husband placed a hand on her bowed head and said, "*Usiogope*! Don't be afraid!"

I was on the verge of changing my mind, of leaving Katito in Africa. I told her as much, but she shook her head vigorously and assured me that it was just a moment of sadness and that it would soon pass. I believed her, and we boarded the plane.

As we took off, Sonkoi, with an intense gaze, waved goodbye and exclaimed, "*Enda salama! Safari njema!* Go in peace and have a good journey!" I have always loved the way Africans wish you a safe journey. It's sweet and kind. Ever since we came to Africa, whenever Luca and I set off on a long safari in the Land Rover, the Maasai would send us off with these words of reassurance. It gave me a deep sense of peace and tranquility. With their blessing in our hearts, we would set off confidently, knowing that nothing could harm us.

We took off peacefully...all except for Katito! The fear and excitement of flying for the first time in her life got the best of her. So, I had to take care not only of my daughter on the plane but also had to deal with her nanny's queasiness.

Once we landed in Nairobi, Katito quickly recovered and confessed to me that despite her discomfort, she had thoroughly enjoyed flying.

The real adventure was still to come. At the airport, she looked around, lost and frightened. She stayed close to me, holding Lucrezia tightly, as if to protect her from some unknown danger. We had made it through baggage drop-off, then the check-in, and finally, we headed to the gate for boarding.

It was then that Katito had her first encounter with an escalator. It took me a good 10 minutes to convince her that it wasn't dangerous to step on. She was reluctant and clearly not very trusting. I saw the same hesitation, mixed with fear and curiosity, in her eyes that I had seen in the eyes of the young Maasai girls who had first come to work for me. It had happened when I taught them to use vacuum cleaners, dishwashers, and washing machines. The spin cycle of the washing machine especially fascinated them for hours. The vacuum cleaner, however, was nicknamed "the devil machine that eats everything." That strange contraption terrified them because it could make anything disappear in an instant. Once, they came to me, distraught and scared, because "the devil machine" had swallowed all the pieces of *Bao* – the ancient African wooden board game – that was on the Indian table in the living room!

Now, I saw the same uncertainty in Katito's eyes. Could these contraptions really be trusted? In the end, she made up her mind. After stumbling and losing her balance several times, she managed to step onto the escalator. Once she reached the top, she sighed with relief and smiled, telling me it had been great fun and asking if we could do it again!

The flight from Nairobi to Zurich was much easier than expected. Katito had gotten used to flying, so she was no longer afraid. We showed her how to fasten her seatbelt and use the bathroom. She was seated next to an African girl, and they chatted amicably throughout the flight.

When we landed in Zurich, after eight hours of flying, Luca and

I were exhausted from taking care of our little Lucrezia for the entire flight. But Katito was as fresh as a daisy, looking around with curiosity and excitement. This was the beginning of her "Italian adventure".

Katito's stay in Italy was full of surprises and new experiences, but unfortunately, there were also teary moments and a sense of longing for her beloved Africa and her family.

I took her to the sea for the first time in her life. Faced with the vast expanse of water, she stood in awe for a few minutes, silently contemplating it. Trying to convince her to take a swim or at least dip her feet in the water was futile. She was adamant. She exclaimed decisively, "You must have a lot of rain in Italy to fill a hole with so much water! Get in there? Are you crazy? How could I walk? I would die immediately! Who knows how many fish are in there and how dangerous they are..."

Katito wasn't afraid of hyenas and lions, but she was terrified of harmless little fish! Nothing I said could reassure or comfort her. Every time Luca and I entered the sea to swim, I could see her silhouette in the distance, standing tall and tense on the horizon, scanning the water for us. She would only relax when we returned, safe and sound, under the umbrella.

One Saturday afternoon, in Forte dei Marmi, Katito and I were walking in the center of town with Lucrezia. Everyone was busy shopping, going in and out of the stores, loaded down with bags and packages. Katito was pushing Lucrezia's stroller beside me. She moved slowly and gracefully next to me. I had always been struck by her innate grace and class. She was certainly no less impressive than the beautiful models who strutted elegantly down the catwalks during fashion shows. She wore her traditional garments, vibrant and joyous. The beads and shells of her Maasai jeweler jingled and sparkled in the sun. The more she felt like she was being watched, the prouder she became, not at all intimidated by the curious glances of the people.

We passed by an Indian man sitting on the side of the road in a wheelchair; he was an old, frail man, blind and crippled, begging for money. No one stopped to help him; no one gave him money or food. We were all (me included) so caught up in our superficial desires and whims that we failed to notice that someone had real needs and was asking for our help.

Katito stopped in front of him; she observed him with compassion and then, with sternness and resolve, asked me, “Why doesn’t anyone help this man? You have money and food; how can you be so indifferent?” She told me that in her Maasai village, no one would leave a needy, sick person on the street without food or shelter; no one with a heart...

I did not know how to answer her questions, but I felt ashamed of myself and of our world. I found myself reflecting on what she had just taught me. It felt as though we were the wild, uncivilized ones.

When I returned to Africa, I asked Katito many questions, curious to know what she thought of Italy. She paused for a moment, then simply said, “In Italy, you have beautiful homes and many cars. Everyone has running water and electricity. You have hospitals and doctors; your children are cared for and most of them don’t die when they are sick, unlike ours! But you no longer have the ‘Things of God.’ The trees, the flowers, the animals... everything in Italy is new and functional, but it wasn’t placed there by God, but by men. Where is God in Italy? Where is nature?”

In Africa, the baobab and the acacia, with their gnarled bark and twisted branches, the vibrant flowers and wild animals – everything is renewed each day in its timeless diversity. Nature is the master, reigning over men, as it always has. And what of nature’s imprint is left in the First World? Have we erased all tracks of it, along with our humanity?

Dear, wise Katito, how much I still have to learn from you...

Meanwhile, in our beloved Africa, something terrible was happening without our knowledge. Something that would forever change our lives and teach me just how harsh and merciless Africa could also be.

The Great Fire

It was the beginning of September 2002.

I was happy to be in Italy with Luca and Lucrezia during this time of year. Since moving to Kenya, September had always been my least favorite month. The heat was stifling, and everything seemed desolate and dry. The dust, fine as talcum powder, would suddenly rise from the ground, carried by the wind that howled endlessly from the Chyulu Hills, creating spirals and whirlwinds. They moved swiftly across the scorching plains, like screaming ghosts that appeared and vanished in an instant, carrying thorns, twigs, and dry leaves.

Over the years in Africa, I had often seen poor, tiny acacia trees bent under the force of the wind, leaning in the opposite direction. And nearby, those that were protected by one of the camp's buildings, grew strong and vigorous. They created a strange contrast to their less fortunate siblings.

Everywhere, the grass was yellow, dry, and sparse, and in some places, where the Maasai herds had overgrazed, a kind of desert had formed. The trees had lost their leaves, and all of nature seemed motionless, as if conserving its last bit of energy, waiting for the longed-for rain and the return to life.

When the rains were near, the plants would begin to sprout tender green leaves, and the bullfrogs, down in the camp's pond where the wild animals drank, would start croaking joyfully, as if to share with us the delight that nature was ready to renew itself. It was something that always amazed and fascinated me. How could the trees and animals "feel" that the rain was coming to bring them back to life? It was a mystery I could not explain, but one that constantly captivated me.

Already in September, the Maasai herders, along with us, would begin scanning the sky hopefully, looking for any sign of rain. Usually, the short rains did not arrive until after the second half of October; sometimes, however, by the end of September, we could see immense grey clouds, with flat bellies, heavy with rain, moving swiftly across the infinite African sky, driven by the incessant wind.

If the rains were near, the Maasai and the Wakamba, a tribe that lived on the eastern side of the Chyulu Hills, would start burning the pastures on the hills. They knew that the imminent rain would make the grass grow greener, and this would be a blessing for both tribes. The Maasai would finally see their cattle regain health after the difficult dry months that had just passed; the Wakamba would be able to harvest the tender shoots of *miraa*, a stimulant chewed mainly by truck drivers to stay awake and drive long distances, to sell in Somalia and Kenya for a few shillings per kilogram. It wasn't much money, but it would allow them to survive.

In the evenings, the perfectly rounded peaks of the Chyulu Hills were framed by fires, and the purple flames would tint the mysterious African sky with fiery oranges and reds. They looked like shimmering crowns of flowers, gracefully placed on the heads of magical fairies living in enchanted forests, advancing into the night, driven by the eastern wind. It was such a unique and unusual spectacle that we all stood entranced, as if we'd been hypnotized, watching the tongues of fire moving toward us.

If only it weren't such a great danger to the camp, I would have serenely kept that incredible vision of nature in my memory forever. But we could not afford to wait and watch. At the first signs of fire, our game scouts – rangers – would immediately be sent out with water and fire beaters to take on the flames and prevent the danger from coming too close to us.

Luca would often join them, helping and coordinating this delicate task. They had always succeeded in their mission and had always extinguished the fires. Until that particular year.

That year, unfortunately, the rains seemed distant, and the clear blue sky, free of clouds, gave no sign of promise. The stifling heat of the African sun covered everything with an ashen haze. Kilimanjaro, whose summit usually stood majestically before us in the clear African sky, was invisible, and everything seemed to tremble in the distant plains.

Unfortunately, we had not cut the grass inside the camp, and the firebreak, which protected it, hadn't been burned. In the previous months, we had been very busy with guests, and the birth of our daughter had distracted us from everything else.

When we left for Italy, we entrusted the camp to our manager, Tom, a young Englishman who had recently arrived in Kenya and at Campi ya Kanzi, and to our pilot, John. Together, they were to ensure that nothing happened to the camp in our absence. We had also left instructions for five of the Maasai employed by us, who were more responsible and attentive than the others, to always stay at the camp while we were gone. Since it was the low season, we didn't have any guests while we were away, so most of the other Maasai were sent back to their villages to rest after a very busy tourist season.

So, in early September, we left for Italy with light hearts, unaware of the tragedy that was about to unfold.

A few days later, on September 13 while we were at the seaside in Forte dei Marmi, we received a phone call from Kenya from a dear Italian friend who lived in Nairobi. He said, "We have a problem. Your camp has burned down. We don't know anything for certain yet, but you need to return to Kenya as soon as possible."

The great fire, which had started in the forest on the top of the Chyulu Hills, had advanced rapidly during the day, driven by the howling wind. It had come dangerously close to the camp; the flames, now too high to be controlled by our Maasai, had jumped the firebreak and started burning the grass inside the camp and some of the buildings. In the kitchen, a gas cylinder had exploded from a spark, setting fire to the dry grass of the roofs of the rest of the nearby buildings.

John was flying guests to Nairobi in our plane, and Tom was far from the camp, overseeing the repair of one of the roads. They couldn't help stopping the fire from advancing. In an instant, we lost the kitchen, the vegetable garden, the food stores, the cutlery and crockery, the linen, the office, and all the electrical equipment, from solar panels to batteries and inverters. Miraculously, Tembo House, the main public area of the camp, and the guest accommodations remained intact. Fortunately, no one was injured.

Basically, the camp's heart had been destroyed. We had also lost all the things we deemed priceless. Since we hadn't yet built our house, all our most precious personal belongings were in the office, where we had a safe. Luca lost the guns his father had left him when he died, and these had great sentimental value, his African diaries, all the slides of his childhood and the trips he had taken before coming to Africa. I lost all the letters from my family that I had received in my first years in Africa and the jewelry my grandmothers had left me when they died; I also lost clothes, shoes, and bags, things that were not so easy to replace in Africa! And then, the photographs of our wedding and of Lucrezia as a newborn, the most beautiful books, and my bridal bouquet. But the greatest pain was the loss of one of our dogs, our beloved Jago. We were told that he had been trapped in the flames in the kitchen and died there.

So, the life we had lived in Africa until that moment was suddenly gone forever, leaving a sense of emptiness, rebellion against this injustice, and great pain and dismay in our hearts. It was not the great economic value of what we had lost that troubled us, but the awareness of having lost irreplaceable sentimental things.

Fortunately, it happened when we were not at the camp. I don't think I would have been able to stay in Africa otherwise.

I cannot and will not describe my state of mind at that moment. Anguish, despair, fear, emptiness, and infinite fatigue are the first words that come to mind. My daughter Lucrezia was not even two months old, and I felt so protective of her and, at the same time, so inadequate. Like most mothers do sometimes. I had decided to breastfeed Lucrezia. After hearing about the fire, my milk dried up from the sadness during my time in Italy. I'm not ashamed to say that my first reaction was not wanting to return to Kenya, but to stay in Italy, safe, with my family to lean on, and able to help me start a new life.

I expressed these doubts and confessed my fears to Luca, but my husband was incredibly strong, determined, and calm, as always in the difficult moments of his life. He reassured me and said, "Don't worry, I'll go ahead. I'll rebuild everything, and then, if and when you feel like it, you can join me with Lucrezia. For now, just think about our baby, and I'll take care of the rest."

And that is what happened. Luca took the first flight to Kenya, where he spent three months living in a small tent to rebuild Campi ya Kanzi. I stayed in Italy with our daughter, trying to fill her days with as much love and peace as possible, while I hid the tears and fears that weighed on my heart.

In Italy, I repurchased everything we needed to bring the camp back to life – plates and cutlery, linens and mattresses, food and wine, electrical supplies and solar panels, freezers and refrigerators. I packed it all into a container, which was sent off to Kenya.

Everyone at the camp worked tirelessly during those three months. Luca most of all, but even the Maasai played a significant role. They worked without pay throughout that time, driven by a desire to help us start anew, understanding that we couldn't afford to compensate them. Their selflessness moved us deeply.

By December 15, 2002, Campi ya Kanzi had been rebuilt, more beautiful than before and ready to welcome its first guests. With Lucrezia, Katito, and my mother by my side, we returned to Africa, ready to restart our lives with Luca and our Maasai community.

With time, Luca and I have come to understand that the most beautiful memories and precious things destroyed in the fire were not lost forever. They are safely tucked away in our hearts, living within us each and every day.

I have come to accept that Africa gives abundantly yet takes away just as suddenly, without warning or reason. Perhaps this is why it teaches us to cherish what we have and to live fully in the present. And it is for this that we love her so deeply. As if we know that what we have today might not be here tomorrow, and so we savor every gift she offers us. But isn't that the allure of Africa? And isn't that the way we should learn to live? Each and every day of our lives?

Oh, how many lessons this wondrous continent has to offer...

Hippos in the Garden

Michael and Sonia were dear friends, companions from the start of our African adventure.

Michael was a striking man – tall and athletic, with blond hair and light eyes. He was charming and captivating, a gallant heart-throb who made women swoon, though he wasn't looking for a relationship. Until he met Sonia. He arrived in Kenya in the early '80s as a diplomat at the Belgian Embassy. When his mandate ended, he invented a way to stay in Africa, a land he, like many others, had come to love by chance and chose to make his home.

Michael built a delightful lodge in Karen, at the foot of the Ngong Hills. His idea was original: instead of a traditional hotel, he constructed wooden treehouses to accommodate guests. The tree-houses were luxurious – spacious, bright, cozy, and tastefully decorated with refined African furnishings. The lodge became the perfect stopover for travelers needing a night's rest in Nairobi, either at the beginning or end of their safari. The atmosphere was relaxing, and the strictly Belgian cuisine was excellent. Over the years, we sent many guests to Michael, and they always returned with glowing reviews.

Sonia was several years younger than Michael. She was beautiful – tall, brunette, with short hair and a face of delicate, sensual features. Her emerald green eyes were wistful and full of dreams. She was an artist in everything she did. Though she was a woman of few words, she had a warm, witty charm. At parties, she never held back, dancing the most and joking with everyone. She often slipped into a world of her own, detached from reality, where her mind spun countless ideas, always unusual and original.

In the early days of her career, Sonia worked in London, designing and crafting costumes for the city's major theaters. She arrived in Kenya in the late '90s, unsure of her path. Her true passion was creating and designing clothes, bags, and jewelry, which she began to do in Nairobi, alongside talented African artisans. Over time, she found fame and success, in Kenya and abroad, becoming an internationally renowned designer.

I shared a special bond with Sonia. Though we didn't see each other often, we shared an important moment in my life – Sonia designed my wedding dress. She was as moved as I was on my wedding day, dressing me and making final touches to my hair and makeup before I joined my husband. Over the years, I bought many pieces from her – refined and sensual suede corsets, elegant cocktail dresses, grand evening gowns, jewelry crafted from precious African stones, and leather bags adorned with Maasai beads. Her style perfectly matched my taste. Whenever I visited her boutique in Nairobi, I knew I'd find something unique to take home.

Sonia met Michael in Nairobi, and it was love at first sight. They were together for a few years before deciding to take the "big step." Finally, after Luca and I married, they decided to do the same, and we were thrilled for them. They chose to wed at sunset, surrounded by their closest friends, on the shore of a small island on Lake Baringo, in northern Kenya.

The flight from Campi ya Kanzi took about an hour and a half. We landed around noon on the shore of the lake, greeted by a gust of hot, dry wind. Lake Baringo, one of Kenya's great volcanic freshwater bodies of water, sits at around 1,000 meters above sea level, its climate dry and arid. It is hugged by tall, barren mountains. To the north, the dormant Korosi volcano looms over the lake, peaceful and solitary. The water's color shifts with the time of day and the sky's hue, ranging from chocolate brown to light beige, from ochre earth tones to precious bronze.

The lake is famous for its vast population of hippos and crocodiles. Large tilapia fish swim in its dark waters, providing sustenance to the fishing communities along its shores. Numerous seabirds, from pelicans to fierce fish eagles, fly low over the water, especially at dawn and dusk. Sometimes, if you're lucky, you might even see delicate flamingos – pink beauties that migrate from nearby Lake Bogoria. When they take flight together, it's a magnificent sight – a pale pink splash against the deep blue sky and the lake's brown waters. Nature's poetry.

We had booked a cottage by the lake to stay the night. It was built of wood and stone, cozy and comfortable. The ground floor had a lovely kitchen and a living room with large windows overlooking the lawn, a perfectly manicured green swath. We were told that at night, hippos would emerge from the water to graze in our garden. I found it hard to believe but hoped to be lucky enough to see them. But the most beautiful room was undoubtedly the bedroom. It was situated upstairs, in a loft, with a large four-poster bed in the center and enormous windows facing the lake. From there, you could see wooden boats with fishermen gliding slowly across the dark waters, hippos – noisy and massive – surfacing now and then for air, Verreaux's eagles and pelicans flying low in search of food, and hornbills and elegant blue starlings singing among the acacia branches in our garden. Occasionally, we would spot crocodiles on the lake's muddy banks, waiting silently for their unlucky prey.

The mother of a dear friend, who had lived by Lake Baringo for years, had been attacked by a crocodile many years before, near our cottage. Every day, the woman swam in the same spot without any trouble from the crocodiles. But that morning, she unknowingly disturbed a young female who had laid eggs too close to her home. The animal, nervous and worried about her offspring, felt threatened and defensive. In an instant, the crocodile tore off her arm. The local fishermen and Flying Doctors responded immediately. Miraculously, they managed to rescue the woman from the crocodile's jaws and save her life! She was flown to Nairobi, and from Wilson Airport, rushed to Nairobi Hospital for surgery that lasted hours. In the end, she survived. She returned to her home on the shores of Lake Baringo and resumed her morning swims in its warm, dark waters. Her harrowing story spread throughout Kenya – few could claim to have survived a crocodile attack! I recalled her tale as I gazed at the lake from our window. I shuddered and began preparing for Michael and Sonia's wedding.

At sunset, two fishermen in a small rowboat came to pick us up in front of the house. After about 10 minutes of crossing, they dropped us on the shores of a small island where the ceremony would take place.

There was a clearing of acacias by the lake. Under the shade of those beautiful trees, an altar had been set up. The bride and groom arrived—radiant, happy, and excited. Sonia's dress was a masterpiece –white, flowing, a bit vintage, and naturally, designed by her. She wore cowboy boots instead of traditional white shoes! The contrast between the elegant, lace gown and the wild, sexy boots delighted me. I thought about how Sonia was always different, unique in everything she did or said.

After the ceremony, we headed by boat to the opposite side of the island. There, a charming lodge awaited us, where the reception was held. The tables were elegantly set poolside. Oil lamps, candles, and white rose petals were scattered everywhere - romantic and exotic. That night, the full moon shone bright. Its silvery light reflected in the vast, dark, and silent lake, deep in slumber. After a delicious dinner and wedding cake, the dancing began. Michael and Sonia loved to dance. The music at their wedding was beautiful. We danced until the wee hours, laughing, joking, and sipping too much champagne.

Just before dawn, we bid the newlyweds farewell with hugs and kisses. On the way back, a sense of peace and mystery enveloped the lake as the fishermen rowed us home. The full moon was still high in the sky, but already, on the horizon, the purple glow of the rising sun could be seen. Luca and I stayed in silence, holding hands, lost in our thoughts until we reached the dock of our cottage.

Back in the bedroom, I undressed and took a shower, still enveloped in the magic of that night. Suddenly, I heard them – grazing steadily, breathing heavily, moving silently below us. I stirred slowly, not wanting to make a sound. I peered out the window, and there they were! So many, too many to count! The hippos were peacefully grazing in the garden of our cottage. They were everywhere. My heart overflowed with emotion to feel them so close, smell their scent, breathe with them...

For this moment alone, it was worth going all the way there, to that remote, primordial place, ancient and serene. At the first light of dawn, they disappeared, one by one, slowly and silently. They slipped back into the lake waters, leaving our hearts filled with wonder and awe. Yet another precious gift from our Africa, to be cherished forever within us.

With My Dogs in Africa

I often think back of one of our dogs: Womble, who had belonged to our pilot.

Womble was a beautiful white Labrador who arrived at Campi ya Kanzi in July 2002, the same month my daughter Lucrezia was born. John had adopted him because his previous owners had to leave Kenya to move to Australia and couldn't take him with them. John had known Womble since he was a puppy and couldn't bear the thought of that sweet dog being left orphaned.

So, one July morning, Womble landed on our airstrip with John and was immediately welcomed with joy and enthusiasm by all of us at Campi ya Kanzi. To know him was to love him.

He was a gentle, kind dog, affectionate and warm. When John was in flight or away from the lodge, Womble would lie patiently under his desk in our office, waiting for his return. He was the exact opposite of Kibo, my own playful and energetic white Labrador.

Perhaps because they were so different, the two dogs became fast friends and inseparable companions. Every day, they would bathe in the pond in front of our tent; they adored the water, like all Labradors, and played and splashed until they were utterly exhausted.

In the afternoons, when the sun's heat began to wane, they would set out on their adventures across the open savannah. Kibo would lead, with Womble always trailing behind, in search of new thrills and excitement. They usually returned at sunset, tired, thirsty, and wagging their tails. But sometimes, they would stay out all night, worrying us as we awaited their return the following morning.

Upon their return, their fur was covered in ticks and prickly wild seeds; acacia thorns often pierced their paws, making them limp for days. They smelled terribly, with breath that reeked of rotting meat. They were always incredibly thirsty, gulping down water from their bowls before collapsing, exhausted, for hours, their half-closed eyes contentedly gazing out over the boundless savannah. It was their private hunting ground, and none of us truly knew what they did during their long absences from camp. I believe that in those moments, our dogs were completely happy and fulfilled.

At Campi ya Kanzi, we had always had many dogs.

The first two, Jago and Sean, had come from Italy. Luca had kept them at his country home, and when we decided to move to Africa, he could not bear to leave them behind. They were, respectively, an Irish Setter and a Gordon Setter; magnificent specimens who quickly adapted to Africa and the freedom it offered. They were happy at Campi ya Kanzi.

I remember their first days in the savannah: they watched zebras and giraffes with curiosity and fear. They regarded the Maasai in much the same way! But soon they were perfectly at home, turning into true African dogs.

Unfortunately, they did not live long. Sean, the Gordon Setter, died just three years after we arrived in Africa. He caught tick fever, and his death was slow and painful. Unlike the other Setter, even though he was no longer a puppy, Sean always seemed to be smiling, forever young.

His passing was a relief for all of us who had been used to seeing him cheerful and lively, and who had watched him, in the end, so sad and ill. It was a great sorrow to be with him in those final days; he died slowly, and the weaker he became, the more affectionate he was. Wherever we went, he followed, slow and limping. The pats and gentle taps on his damp nose were never enough; and this was a dog who had never wanted to be petted! It was as if he knew he only had a few days left with us and wanted to enjoy them fully before leaving us forever.

When Sean left us, I felt profound sorrow and went through a very difficult time. To console me after his death, they gave me Kibo, a whirlwind of a Labrador puppy who certainly filled my days with joy and helped me recover from the loss of my dear friend Sean (though no dog can ever take another's place in my heart).

Unfortunately, not long after, the terrible bush fire of September 2002 also took my beloved Jago. Jago had always held a special place in our hearts. His father, Toby, had been the best dog Luca had ever had. An exceptional hunting dog, in Italy, Toby had been Luca's shadow. Wherever Luca went, Toby was with him. At times, I think he had his own feelings and thoughts. His deep brown eyes would gaze at you for a long time, and if you were willing to listen, they could speak to you.

Toby died while we were still in Italy, and Luca took the loss very hard.

Jago was not as unconditionally loyal as his father, but he was still an extraordinary dog. He was elegant and proud; his long, soft, russet fur shone in the sunlight, reminding me of the autumn-hued avenues of my beloved Italy. I loved his personality: he was independent and solitary yet knew how to be sweet and affectionate. Just like me. He was quick-tempered and irritable at times, but he was immensely patient with Kibo, who, as a puppy, never gave him a moment's peace! He tolerated the little Labrador's disrespectful antics with patience and resignation, enduring Kibo's tugs at his long, soft ears until his fur was nearly pulled out.

Jago was intelligent. He understood how important Lucrezia and I were to Luca, and after our daughter was born, he became an incredible guard dog. He protected us bravely and kept us company in our moments of solitude. He was truly an important presence in our lives and a dear, trusted friend to me. At night, he slept with us in our tent, and his gentle breathing gave me a sense of tranquility and security. Sometimes, I would feel him stirring in his sleep at the foot of my bed; perhaps he was happily chasing guinea fowl and francolins in his dreams across the endless plains.

One morning, I saved him from a lion attack. I don't know where I found the courage, but I remember seeing a lion in the tall grass, leaping towards Jago, and I screamed and charged at it, furiously shooing it away! Probably, the lion was more frightened than I was. My screams had the desired effect, and the lion abandoned its pursuit of Jago and disappeared back into the savannah from where it had come. Shortly afterward, I found myself trembling with fear, reflecting on what had just happened. But Jago was safe, and that was all that mattered.

I had hoped to teach my daughter Lucrezia to walk, leaning on the soft back of sweet Jago. But fate had other plans and took him from us too soon, far too soon for our weary hearts.

When the big fire broke out, Luca and I were in Italy. The Maasai told us that Jago was trapped in the kitchen, surrounded by flames. And there he died. I will never forgive myself for not being there with him when he passed. What did his eyes see before closing forever? Not our faces, as it should have been. Did he think we had abandoned him? I will always have regret in my heart.

Every time I lose a dog, I am filled with deep sorrow; I vow never to get another. I grow so attached to them that they become true friends, essential presences in my life. The fact that they are dogs, not people, makes no difference. At times, I believe I have suffered more from the loss of my dogs than from the loss of people who didn't play as significant a role in my life.

But then, whenever I see a puppy, I want to take it with me. And so, another dog becomes part of our large family at Campi ya Kanzi; I try to fully enjoy its company while it is alive and with us. I have accepted that having a dog in Africa means being ready to lose it at any moment due to an inexplicable and incurable illness, an aggressive lion, a leopard, or the fatal bite of a snake. Despite this, I have decided that I cannot live my life without a dog by my side, and I am therefore willing to risk suffering. It is always worth it.

If I am able to have so many dogs, I owe it to Patrick, my cook, who, with his patience, dedication, and deep love for animals, helps us care for my faithful friends every day, ensuring they lack for nothing.

When the weight of losing my dogs becomes too heavy for my heart, I think of an endless savannah under the fiery sky, in the warm light of the setting sun. There, among the long golden blades of grass, I see Toby, Sean, and Jago running happily, tails wagging, sniffing the air for a familiar scent carried by the wind... now, Womble and Kibo are there too, following them from afar. Soon, they will join them.

When the time comes, I will be there with them, and it is comforting to know that I will see them again, that I will once more be able to stroke their damp noses and long ears and find the same sweetness as ever in their kind eyes. Then, with my faithful and loyal friends, at sunset, I too will walk the infinite plains, just as we did long ago when I didn't yet know how much I would miss them...

African Nights

I love the nights in Africa: they are the most beautiful moments of my life here, the sweetest and calmest. The nights are my "friends," and I have never felt alone or frightened while immersed in their mystery. When I first arrived in Kenya, while we were still living in a small tent, I would often wake in the dead of night, startled by the absolute silence that surrounded me. It was almost painful to my ears, pressing down on my heart.

And yet, it was in those precious moments that my deepest and truest thoughts came to be. I would think about where I was, in the midst of a wild land, with no one, except Luca, beside me, completely immersed in the purest and most authentic nature. I was a tiny dot lost in the immensity of the universe: and it was then that I felt the full grandeur of Mother Nature and was profoundly grateful for my life and its mystery.

Setting aside the initial fear and the anxieties of my former world, I would surrender to this sweet sensation of solitude, let go of everything, and my heart would rest.

After my car accident in Italy, I had gotten into the habit of waking two or three times a night. And I actually liked it. Most of my writings came about in those moments of absolute silence and profound peace. Only in those moments could I listen to my heart and find within it the things that were secret and true.

As soon as I woke up, I would strain my ears, searching for a sound or a rustle that would bring me back to reality. But nothing... I liked that feeling. It was like being suspended in the silence, in a bubble of air, above all things, in a dimension without time or place. Immersed in complete darkness, I would think about how the first humans on Earth must have felt: small and insignificant before the immensity of it all, yet also powerful, because they were at the center of the universe. And that's how I felt, in those beautiful dark and silent nights.

Yet, if I listened closely, I would soon realize that I was not alone: the voice of the wind, sometimes almost a whisper, spoke to me, and the crickets and frogs at the pond sang along with it. The ancient breath of the Earth, Mother Africa, was in the background, reassuring me. The flickering light of the endless stars hanging up there in the black African sky illuminated the night, and the silhouette of Mount Kilimanjaro was always there, like an old, wise father watching over his beloved child.

And the moon, what can I say about the immense African moon? Light, magic, enchantment, mystery, precious silver... Who knows why the sun – which, after all, represents life – didn't evoke the same deep emotions within me? The same passion, the same poignant sense of melancholy...

When there was a full moon, we liked to sleep with the tent open, to let its magical light flood over us and to steal some of its precious energy. Our tent stood in front of a small pond. Each night, the ritual of the savannah creatures coming to drink would take place.

They always arrived in the same order, and we never witnessed any squabbles or fights over who had the right to drink first. It was as if nature had its own specific ritual, always the same, which all the animals diligently and precisely followed.

Usually, the buffaloes would start, those immense dark animals with curved horns and menacing eyes. They always filled me with a bit of fear. They were enormous and heavy. I had learned to get used to the sound of their hooves on the rocks outside our tent. Sometimes, after drinking, they would linger to graze on the tender grass outside our shelter for hours. We could hear their breathing and smell their acrid scent through the fabric walls. They were less than a meter away from us. It was a unique sensation, of symbiosis and connection with those splendid animals. A sensation that we no longer experience now that we live in a house. And we miss it dearly.

Then along came the graceful gazelles, zebras, and giraffes. The latter were, to me, the ladies of the savannah. They arrived walking with long strides, elegant and light, almost on tiptoe. They always stayed at the pond to drink for just a short time. As soon as they were done, they would leave in perfect silence, as if not wanting to disturb our sleep! If that wasn't perfect etiquette, then what could be?

The noisiest and most boisterous, with no respect for the presence of humans or other animals, were the lions. After all, they were the kings of the savannah, and with their arrogance and regal nature, they liked to remind us of their title. They would arrive late at night, usually after all the other animals had left, and sometimes they would stay at the pond for hours, even just to cool off or rest beside the water. I remember one morning, at dawn, we were awakened by a noise. We slowly unzipped our tent to witness a magnificent sight. There were two lionesses with four cubs at the pond. The little ones were happily playing near the water, and their mothers patiently waited for them to finish having fun. We watched them in silence for a good half hour, enchanted by their beauty.

Over the years, those two lionesses returned many times to the camp. One of their cubs, a female, grew up and often came to drink at the pond in front of our tent. She had become part of our family. We had named her *Nosero*. Her name, in Maasai, means "dense bush," after her chosen shelter in the thick forest of the lava flow next to our airstrip.

She wasn't easy to find, and we didn't dare venture into the forest to search for her and disturb her. She would show herself when she wanted. She was beautiful, proud, and strong, and over the years, she had had several litters. Sometimes, when her cubs were born, and she saw our Land Rovers parked next to the forest, she would come out of the lava with them. She would crouch in the grass of the plains, and, always staying alert, she would allow us to admire her cubs playing with her for a few minutes. It was always a great emotion, and we were grateful to her for trusting us. We had a special relationship with her. She had become the mascot of Campi ya Kanzi.

Nosero could also be mischievous: one evening, while with our guests at Tembo House (the main area of Campi ya Kanzi), we waited for an hour for the food to arrive. Nosero, along with one of her companions, had decided to crouch right outside the kitchen, preventing the waiters from coming out to bring us our dinner. That evening we ate late, and the pasta was overcooked and cold; but for some reason, no one complained!

Often, two different prides of lions would come to drink on the same night. To mark their territory and show their supremacy over the others, they would start roaring in turns. Their roars grew louder and deeper and could go on for more than an hour, preventing us from getting back to sleep.

An old Maasai proverb goes: "*The warrior fears the lion three times: the first time, when he hears its roar; the second, when he sees its tracks; and the last, when he faces it for the first time.*" I can confidently say that what the Maasai say is true. Every time we heard the roar of a lion near us, we felt it profoundly in our stomachs, deep within our very being. And the primal fear of these formidable beasts, felt who knows how many times by our ancestors huddled in a cold cave on the lonely nights of prehistoric times, would come knocking at the door of our hearts, entering with force whether we liked it or not.

One night, I was awakened by the cries of a jackal. They were sharp and persistent, repeating insistently every few minutes. They sounded to me like a plea or a cry for help. So, I quietly slipped out of my tent and turned on the flashlight in the direction of the cries. I spotted a small jackal, crouched in the grass, not far from the pond. It was looking toward the waterhole and kept whining. Peering into the darkness, I finally understood the reason it was so agitated: an enormous lion was crouched beside the pond, blissfully enjoying the coolness of the water in the night's heat. The jackal was probably very thirsty and hoped to convince the lion to leave and let it drink. But there was nothing to be done. The poor little creature continued its strange litany for almost two hours without success. Only at dawn did the jackal quiet down and give up on drinking, leaving disappointed and thirsty. Once again, precedence had gone to the lion, the proud and undisputed king of the savannah!

Mischievous Lions and Hyenas

The hyenas, jackals, and lions could be quite mischievous. Shortly after Campi ya Kanzi was completed, in the mornings, we often found the hammock hanging under the grand fig tree gnawed on by hyenas and the cushions on the sofas chewed up by jackals.

One morning, Luca was upset with me because he couldn't find his shoes. He had left them outside our tent the night before, and now they were gone. He thought I had hidden them as a prank. But it wasn't me; the hyenas had played the joke, sneaking away with his shoes during the night, only to drop them a short distance from the tent, with the laces all chewed up. We found them a few days later, with remnants of leather strewn along the grass.

One time some lions decided to play with the cowhide poufs that were arranged by the fireplace. Perhaps they thought they were calves and wanted to claim them as their own! The next morning, we started searching for the poufs everywhere. We finally found them a few hours later, a couple of hundred meters from Tembo House, lying in the grass beneath an acacia tree. Aside from a few scratches and bite marks, fortunately, they were intact. The lions did not find them particularly appetizing and had abandoned them after playing with them for a short while.

From that day on, we decided to station *askari* (guards in Swahili) to watch over Tembo House and the other buildings of Campi ya Kanzi every night, ensuring that such incidents wouldn't happen again.

Our first choice wasn't the best. Paramoja, the Maasai we assigned to guard the camp, was a quiet and serious man. He stood tall and proud in his fiery red *shuka*, with his long spear always resting on his shoulder. He seemed the perfect fit for the job.

As time passed, we noticed that Paramoja was never alone, and whenever he had to patrol the camp to check for nearby animals, he would ask another Maasai to accompany him. Finally, we hesitantly asked him why he always needed someone with him. Paramoja's face darkened, and without a word, he lifted his *shuka* to reveal a scar running the entire length of his thigh.

He explained that when he was a young man, participating in a lion hunt with his circumcision group, an enormous lion had attacked and nearly killed him. That scar was the mark left by the beast. Naturally, after that terrifying experience, Paramoja was deathly afraid of lions, so whenever he had to be alone, he wanted another Maasai with him, acting as his bodyguard. From that day on, Paramoja was assigned to a different role within the camp. No one ever spoke again of what he had shared with us; I believe he was immensely grateful for our understanding and for helping him maintain his dignity in front of the other Maasai working at the camp.

After Paramoja, we opted for two Maasai who couldn't have been more different, but together they formed a winning team. Their names were Daudi and Munghesia. With them around, Tembo House's furnishings were safe!

Daudi was very tall and slender as a reed. He was over 40, married to a girl 20 years his junior, and had two children. He loved his job and took great pride in it. He was a man of few words, always serious, respectful, and composed. I rarely saw him laugh. With me, he always spoke only Maasai and a few words of Kiswahili. Later, I was told by American guests that he also spoke English, as he had had some very interesting conversations with them. But he never did with me! Munghesia was shorter and a bit stocky. Younger than Daudi, and his complete opposite in terms of personality and demeanor. He was always cheerful and smiling, always ready for a joke or a playful remark. He loved chatting with our guests, was curious, and always good-natured. Together, they were perfect, balancing each other out. I had never seen them argue or show disrespect toward one another.

Daudi touched my heart with what he did when my mother and Luca's mother came to the camp. My daughter Lucrezia had just been born, and the two Italian grandmothers had come to help me care for her. They stayed for a month, sharing one of the guest tents. Without us asking, Daudi stayed outside on the veranda of their tent every night to make sure nothing happened to them. His desire to watch over Lucrezia's grandmothers moved me deeply, as if he were their guardian angel.

On the day of their departure, Daudi arrived with two live chickens as farewell gifts for the "*shosho ya Lulu*" (Lucrezia's grandmothers, in Kiswahili). The cost of those gifts represented a significant economic sacrifice for him. I was speechless at his generosity and kindness.

The grandmothers were very flattered by the gifts but explained to him that they couldn't take them to Italy as they wouldn't be allowed on the plane. Daudi was perplexed and silent for a moment. Then he exclaimed, "*Hakuna matata!* (no problem!). I will keep the chickens at my village, take care of them, and bring their eggs to Antonella every day so she can save them for when you return to visit us!" Sweet, gentle Daudi! We were all touched and thanked him from the bottom of our hearts for his kind gesture.

Even now, years later, Daudi still brings me the eggs from the *shosho ya Lulu*'s chickens every week, and heaven help us if one breaks on the way from his village to Campi ya Kanzi! And where else but in Africa could something so beautiful and touching happen?

Leshan

At Campi ya Kanzi, time flew by swiftly; days turned into months, and months into years. Lucrezia was growing up, perhaps too quickly. We followed her every step with wonder, pride, and satisfaction, overflowing with gratitude each day for the precious gift we had been given.

Our daughter was surrounded by boundless love – from us and from the Maasai who lived with us. She was serene and joyful, sensitive and sweet, sharp and intelligent. She was also mischievous and spirited, loved playing pranks on everyone, and was curious and eager to learn something new every day. She adored classical music and claimed she would become a ballerina when she grew up. Naturally, we made sure she had a pink tutu and a pair of ballet slippers. On sunny, breezy African afternoons, we would often see Lucrezia dancing lightly and gracefully around her favorite acacia tree, as delicate as a butterfly. It was such a sweet and moving sight: a little blonde girl with curls, twirling in a pink tutu and ballet shoes, in the wild savannah, as if it were the most natural thing in the world, happy and unaware of how extraordinary her life already was!

When she was almost four, we began to think about giving her a little brother or sister. Luca pushed for it more than I did, constantly repeating how important it was for our daughter to have companionship in the solitude of her African life.

On July 11, 2006, we decided to host a dinner at the camp to celebrate my birthday. We invited friends from Italy and Kenya. It would be a chance to reconnect with old friends who had been absent from our lives for too long. At sunset, we gathered for an aperitif around a crackling fire under the Longido rock, with Maasai warriors singing and jumping in the golden light of the fading sun. Then we returned to the lodge, singing loudly on the Land Rover, already a bit tipsy and excited. After a warm shower, we dressed for dinner: the ladies in evening gowns and the men in tuxedos. Arriving at Tembo House, we sat down to enjoy fresh lobster and crab (that had just arrived from Mombasa on our plane), generously accompanied by our Franciacorta sparkling wine. The delicious dinner was followed by dancing under the stars and much laughter until dawn.

The next day was Lucrezia's birthday. We celebrated it with the children of our friends and my daughter's Maasai friends; Patrick made her a cake shaped like a giraffe, her favorite animal. Gifts, colorful balloons, paper hats, streamers, confetti, and outdoor games made the day truly unforgettable for the children.

I received many gifts from my friends, but the best birthday present was undoubtedly discovering I was pregnant again. Just as it had happened with Lucrezia, I was certain of having conceived another baby in the magic of that July night.

Unlike my pregnancy with Lucrezia, this one was difficult. I had problems from the start, and the nausea and discomfort haunted me until the fourth month had passed. I was constantly exhausted and sluggish. I did not have the incredible energy I remembered from when I was expecting Lucrezia, nor the sensation of being at the center of the world and invincible that had characterized my first pregnancy. I remember telling Luca: "It must be a boy. I can feel it. I feel so different from when I was expecting Lucrezia."

By mutual agreement, we decided, as we had with Lucrezia, not to find out the baby's sex until birth. We wanted it to be a surprise. But I was sure throughout the pregnancy that I was carrying a boy. And I was right. We named him Jacopo.

Jacopo was also born at the Nairobi Hospital. It was two in the morning on April 3, 2007. The same gynecologist and pediatrician who had assisted with Lucrezia were with me. The birth went well. Luca was by my side, loving and patient, as he had been when Lucrezia was born, encouraging me during the toughest moments, massaging my back, breathing with me and holding my hand between contractions.

Like his sister, Jacopo was a beautiful newborn. He was 51 cm long and weighed 3.7 kilograms. His enormous green eyes only showed their color a few hours after he was born. Perhaps he didn't want to see what surrounded him now, having been so content and happy in the womb. As soon as he was out of my belly, they placed him on my chest, and he immediately latched on to my breast, sucking intensely. I felt very well cared for at the hospital, as usual. All the doctors and nurses remembered me. They were professional and very affectionate with me and the little one, and a week after his birth, we flew back to the camp, eager to introduce the Maasai to the new arrival.

When we arrived, all the Maasai were lined up in front of Tembo House, dressed in their finest clothes and wearing their most colorful jewelry to welcome the *moran* (meaning warrior in Maa language) of Campi ya Kanzi. They brought him small gifts and beaded bracelets. Just as with Lucrezia (*Namunyak*), they decided to give Jacopo a Maasai name: they called him *Leshan*, He Who Brings the Rain. Indeed, we had been surprisingly fortunate since Jacopo was born. Abundant rain had poured over the camp after long months of drought, breathing life back into the parched land and filling us all with great joy. Kilimanjaro stood snow-capped against the blue sky, and everything was beginning to turn green and alive once more.

The first night at the camp, Jacopo, like his sister, received a visit from a pride of lions who, roaring repeatedly outside his bedroom window, welcomed him. The little one didn't even stir in his cradle. He continued to sleep peacefully, as if the roaring of lions were the most natural sound in the world! The next evening, at sunset, while I was nursing him, I spotted a lion and lioness mating just a few steps from his bedroom window, wild and tender at the same time... what a feeling, what a unique experience I was living, what a privilege to witness such a sight. Once again, I had the sensation of living a unique and unrepeatable moment, to be cherished in my heart.

That night, Luca took Lucrezia outside onto the veranda to show her the lions shining a flashlight on them. She was not at all frightened and stood mesmerized, watching them wide-eyed, and when her father told her to come back inside so as not to disturb the lions, she replied without hesitation, "You go inside, baba, I'm staying out here with my lion friends..."

Thankfully, my mother and mother-in-law were with me at the camp once again to help me during the first month after Jacopo's birth. Lucrezia was very jealous at first. Luca and I gave her more attention and affection than usual, but it never was enough. Then, slowly, she got used to the presence of her little brother, accepted him, and became sweet and protective towards him. Watching them play together, witnessing the growing love for one another in their innocent eyes, filled my heart with joy and made up for all the difficulties and incredible fatigue of that initial period.

I had decided to breastfeed Jacopo, but I had not anticipated how insatiable and different from his sister he would be. My son wanted to nurse every two hours or so (day and night), and each time he would drain my breast of every last drop of milk. After each feeding, I felt empty and exhausted. The love I had for my son and the sight of him growing strong pushed me past the obstacles of that exhausting time. I nursed him for six months, then finally, with relief but difficulty, I was able to introduce him to formula in a bottle and gradually wean him.

Katito, the same Maasai woman who had raised my daughter Lucrezia with me, was adamant that she wanted to care for Jacopo as well. Lucrezia was now older and no longer needed her. She was absolutely convinced that she was the best of all the women at the camp to look after the new arrival. I agreed, but neither of us had accounted for the fact that Jacopo, unlike his sister, didn't sleep at night. So, we had to find another Maasai girl to help us during his sleepless nights, to give Katito and I the chance to rest. My son Jacopo was indeed a true angel: cheerful and sweet, but only during the day. At night, unfortunately, he refused to sleep, making it incredibly challenging for us!

I had resumed working full-time at the lodge and was exhausted by evening. I could not afford to be awake at night. Lucrezia was becoming more demanding of my time, and I had to devote all my free time to her. If Katito and I were to avoid complete collapse, we had to give in and have someone else take care of Jacopo at night. So, after meeting several women, we chose a sweet and amiable Maasai girl: Elizabeth. Fortunately, she quickly endeared herself to Jacopo and learned to care for him during his long nights of crying and playing.

In October, when Jacopo turned six months old, we left for Italy so our families could meet him. Everyone was eager to get to know the latest addition: they spoiled him incredibly, just as they had, and still did, with Lucrezia.

Jacopo was welcomed by his Italian family with the same great love and tenderness reserved for Lucrezia.

He was a child who was full of joy, always smiling, and even though he was so small, he knew how to assert himself with his older sister. He would poke her in the eye or pull her hair without a second thought if she refused to give him what he wanted, but a moment later, he would seek forgiveness with a shower of kisses and tender caresses.

Lucrezia was more like me and had always been a bit shy and reserved in showing her feelings. Jacopo, on the other hand, was affectionate and outgoing, impulsive and in need of physical touch, just like his father. He definitely knew how to win people over and quickly captured the hearts of everyone around him. Constantly on the move, tireless and endlessly curious, he never stopped exploring his surroundings. Lucrezia would watch her brother in silence, her gaze thoughtful; at times, I could see a flicker of jealousy and anger in her eyes towards this little being who had suddenly arrived in our world yet had already claimed our hearts so completely. So, I would pull her onto my lap, shower her with affection, and reassure her of how deeply we loved her and how important she was to us. I knew she would understand and learn to love her brother, for her young heart was already full of kindness and sweetness. She just needed time.

We stayed in Italy for several weeks, and my children were sick the whole time. The flu, bronchitis, asthma, ear infections... it was an absolute disaster! The fact that my children were born in Kenya and were accustomed to the pure, untainted air of the Chyulu Hills undoubtedly made them more vulnerable than Italian children to every kind of virus or germ. After yet another round of bronchitis, the Italian doctors said, “It’s better for these children to return to where they were born, where they will thrive – in Africa!”

So, in January, we returned to Kenya. To my great relief, within just a week of our arrival, the children had fully recovered. They were eating and sleeping well again. Lucrezia was thrilled to return to the routine of her school and to reunite with her Maasai friends, and a healthy pink color returned to Jacopo’s face instantly.

This made me reflect on the fact that Kenya was my children’s home, the place where they were well and happy. Africa had become an integral part of our lives. And it always would be. It was time to accept this and perhaps begin to feel more Kenyan than Italian.

The Buffalo's Last Breath

A shot rang out, sharp and still, puncturing the silence of the African dawn: my heart stopped for a moment, and I stood tense, alert, listening...

Even Jacopo noticed what had happened and stopped nursing at my breast. I looked out the window. It was a beautiful morning, with Kilimanjaro covered in snow, its silhouette outlined against a sky of pink and orange. The doves began their song, so dear to my heart; nature was awakening, and everything seemed fresh and clean, ready for a new day.

My thoughts immediately turned to Luca. I knew he had been running into a large buffalo on our path for several days and was worried it might charge. The enormous animal was injured in one eye, old and solitary, and therefore, extremely dangerous. For days, it had approached the camp with unsettling ease, no longer fearful; it came as close as the kitchen and our home far too often. One of our Maasai waiters had told us he had been charged by the huge animal with menacing eyes more than once at dusk and had miraculously escaped.

Buffaloes, along with hippos, are the most dangerous animals in Africa. If one intrudes on their personal space, they can charge at a speed that's unimaginable. Many fall victim to buffaloes they never see, hidden in the darkness of night or behind a bush in the midday heat. One rarely survives when charged by a buffalo. Therefore, like elephant hunting, buffalo hunting is extremely dangerous. A buffalo, when wounded, will lie in wait for the hunter who shot him, charging and piercing him at the right moment, to finish him off without mercy. We had always held great respect for these immense animals. We knew there was never a second chance if a buffalo decided to come at you.

Often, in the evenings, walking home after dinner from Tembo House, we had come upon buffaloes grazing right on the path to our home. We would wait for a few minutes, shining our torches on the enormous animals in complete silence; always alert and ready to flee. If the buffaloes didn't move, Luca and I were forced to turn back and fetch one of our Land Rovers to drive home. The right of way always belonged to them on the paths of the savannah!

Luca had left the house that morning with his rifle on his shoulder. He told me later that he had a premonition, a feeling that something terrible was about to happen. My husband had not fired a shot since we had arrived in Africa. He had been a hunter when he was younger in Italy and later in Tanzania, but after settling in Kenya, he could no longer bring himself to take a creature's life. His hunter's need to possess nature had been overcome by the simple awe of contemplating it instead. Since arriving in Kenya, Luca had become a passionate and conscious conservationist, dedicating himself daily to protecting the African fauna and flora surrounding him. The last thing he wanted to do was pick up a rifle and have to shoot an animal. But that day, unfortunately, things did not go as we would have wanted.

After hearing the shot, I immediately left the house with Jacopo in my arms and ran toward Luca. He stood a few meters away: motionless, empty, and miserable. Beside him, collapsed on the ground, lay the old buffalo that had been approaching us for days. It was crouched, its head resting on its front legs, perfectly straight. It seemed to be sleeping. Its horns were enormous, seamlessly symmetrical. Luca had fired a perfect shot, straight to the heart. The old beast's only fault was having gotten too comfortable being around humans; it had become, therefore, very dangerous for us. If it had stayed away from the camp, it would have died naturally, not long after. Unfortunately, it had crossed our path, and we had to shoot it to prevent it from killing one of us.

Luca told me he had been charged by the buffalo that morning and was forced to shoot. "I didn't want to shoot it," he said. "I tried to scare it away, but it wouldn't budge. It looked me straight in the eyes and came at me!"

I looked at my husband and saw great sadness in his eyes. He had no choice, it's true, but the fact remained that he had been forced to kill an animal, and it pained him deeply.

I touched the buffalo's head, its horns. I could smell its wild, acrid scent, strong and overwhelming. It seemed impossible to me that it was no longer breathing. The infected eye was full of pus, dirty and foul; flies buzzed all around, landing on it, covering it entirely. We walked back to the house, hand in hand, with heavy hearts.

In the afternoon, we made sure the buffalo's carcass was removed and taken far from the camp. The animal had died right on the path leading to our home. The lions, hyenas, and jackals, the scavengers of the bush, would arrive quickly to feast on the poor buffalo's body. That was the sad law of the savannah from which no living being could escape. We could not risk having lions and hyenas outside our house for days, eating and not wanting to be disturbed (understandably). This could become very dangerous, especially for anyone who would have to walk the path, finding themselves just a few steps from the hungry predators.

By sunset, the Maasai had already removed the buffalo's carcass. Only a few drops of blood on the green grass, flattened by the immense weight of the dead animal, remained as a reminder of what had happened early that morning. Africa had returned to its quiet, serene self, ready to begin anew.

That night, we were awakened by low moans, hooves trampling the earth and rocks, small bushes breaking and brushing against the rough skin and under the feet of the animals, heavy breaths, and snorts of disdain. It was a large herd of buffaloes: the brothers and sisters, the sons and daughters of the fallen beast, coming to mourn his death. They searched for his body, sniffed the ground insistently, and licked the blood of their lost friend.

We stayed awake all night, watching what was happening, feeling guilty for what had occurred, and regretting what we had done.

For days and days, the buffaloes continued with their lament. In the place where the large animal had been killed, two or three buffaloes always stood as if guarding the invisible body of their lost companion. They had become much more aggressive than usual: as if they wanted to make us understand how hurt and offended they were by what we had done.

Samson, one of the Maasai who worked with us, had told us that Maasai cows mourned for days, mooing incessantly, standing at the place where one of their own had been killed. Luca and I didn't know this: it made the buffalo's death even more dramatic and wretched.

We had learned another lesson from "Mother Africa." The pain and regret for what we had done would stay with us for many days. We were sure we had done the right and necessary thing. But deep in my heart, the image of that lifeless buffalo, lying on the bare African earth, on that morning of sun and dew, would remain forever. For a moment, I thought I saw a tear escape from the diseased eye of the enormous beast...

Safari in the North

Jake and Aisha were new friends, yet the bond we shared, the deep affection and understanding that connected us, made it feel as though we had known them for many years.

Jake was a famous American actor who was brimming with talent and widely admired. He was my age and had received numerous awards and a great deal of recognition for his performances. Acting brought him joy, but it wasn't the most important thing in his life. Raised in a family of intellectuals, Jake's love for the stage may have been nurtured by his mother, who, as a teacher, took him to the theater from a very young age. His father, a brilliant lawyer, was deeply involved in environmental protection. It was only natural that Jake, too, was driven by a love for nature and a fervent desire to preserve the natural world, endangered animals, and threatened communities.

Years before, Jake had visited Campi ya Kanzi, where he first witnessed our work. From that moment, he became deeply involved in our projects and joined our Maasai Wilderness Conservation Trust, eager to help the Maasai and protect their land.

On the surface, Jake seemed shy and introverted. Highly intelligent, he took his time opening up to others. Once he did, he became unexpectedly talkative and funny, a quality that reminded me of myself.

Aisha, his soulmate, was quite different from Jake.

The daughter of true Canadian hippies, Aisha was raised in a free and informal environment in which expressing one's feelings and showing affection were the most natural things in the world. She was spontaneous, simple, and extroverted, often inhabiting a world of her own, a place of light, space, silence, and meditation. She practiced yoga daily, and when she was at Campi ya Kanzi, we would do it together – just her, Lucrezia, and I. I cherished those sunset yoga sessions on the terrace of our home, breathing in the peace and positive energy of Africa. Yet Aisha was also highly skilled in her work and very practical when needed—she was a Hollywood film producer, one of the best.

I liked Aisha from the moment I met her. We got along well, sharing the same tastes and passions. Both of us were quieter and less adventurous than our partners. So, often, when Luca and Jake went on challenging hikes in the Chyulu Hills or decided to fly at dawn around the snow-capped peaks of Kilimanjaro, Aisha and I stayed at the camp, enjoying each other's company and chatting pleasantly until our loved ones returned.

To my eyes, Jake and Aisha were a beautiful couple, deeply in sync and perfectly balanced. They treated each other with respect and shared a strong bond. Being with them reminded Luca and me of the significance of small gestures and thoughtful acts toward one another – something we sometimes forgot after 20 years together. Rediscovering and reliving those values was a joy.

Jake and Luca were kindred spirits. They shared the same thirst for adventure, the same love of freedom and nature, the same sense of humor, and the same deep honesty. They also shared a great desire to help the Maasai and preserve Africa and its wildlife.

Like Luca, Jake was a pilot. Another thing they had in common: a love for flying over the boundless African skies.

Jake had visited Campi ya Kanzi many times over the years. It had become a sort of second home for him, a base for his travels around Kenya. He had visited Africa several times before. He told us one day that, as a child, one of his greatest wishes was to visit Lake Turkana, in northern Kenya. He finally decided to go there with Aisha and asked us to join them.

Luca had always wanted to visit Lake Turkana, where his father had been with his mother many years before, treasuring beautiful memories of those distant days. He often told me how much he wanted me to accompany him there, certain that I would love Turkana as well.

One evening, we found ourselves around the campfire – and almost without realizing it – planning our future journey with Jake and Aisha to Lake Turkana. The boys were excited and eager to go and show it to us. They were ready for the big adventure!

We would take our plane, a six-seater Cessna 206, and fly from Campi ya Kanzi, over the Rift Valley, to the northwestern part of Kenya, eventually landing at Lake Turkana in Ileret. We would spend a couple of days as guests of the Leakey family. From there, we would fly over the Chalbi Desert and land in a tranquil oasis, Kalacha. Finally, we would fly to the Samburu Park to stay in a new lodge owned by an Italian friend.

Aisha and I were less daring and fearless than our travel companions and were feeling a bit anxious. We were heading to a wild and isolated place with intense heat, well-known for its giant crocodiles, snakes, and dangerous scorpions. Turkana was in the far north of Kenya. To get there, we would have to fly for many hours, something that didn't thrill us as we weren't that fond of flying. I was especially unsure about going. I did not want to leave the kids alone at the camp for six days.

It was not easy to convince me to go, but in the end, I overcame my anxious motherly fears and decided to join the adventure. I would leave the children with Katito, Nareio, and Namunyak, the best Maasai nannies I could find. Clare, Lucrezia's American teacher, would also be at the camp, ready for any emergency. I knew my children couldn't be in better hands.

We took off from Campi ya Kanzi on a cold, cloudy July morning at dawn. The flight from the camp to Lake Turkana would take several hours, so we decided to break up the journey and stop for breakfast at Lewa Downs. Shortly after, we resumed our journey, gliding over a soft bed of white clouds. After almost two hours of flying, Luca turned to us girls, with that strange light in his eyes that always appeared during moments of great emotion, and exclaimed, "We're here!"

Flying over Lake Turkana was a profound experience for all of us.

We silently admired the vast expanse of water, shifting from deep emerald green to shades of blue and turquoise. It appeared like a precious gemstone set in a desert of dark grey rock and sand, making the unusual and intense color of the water stand out even more. It was also known as the Jade Sea, and now we understood why.

Turkana, the world's largest permanent desert lake, ancient and majestic, spread out beneath us in all its enormity and magic. Its surface was rippled by the strong wind that constantly blew across its waters from the surrounding desert. It was home to hippos, large catfish, Nile perch, tilapia, and enormous crocodiles. A wide variety of colorful birds, both small and large, flew close to the water, eagerly searching for a meal.

In the middle of the lake, rose two large, dark islands, which were round and squat, as well as being ancient and prehistoric like the lake itself. As I looked at them, they seemed like old sea turtles resting peacefully before diving back into the vastness of the ancient lake. Near the largest and oldest island, Central Island, many of these prehistoric reptiles truly lived, and it wasn't uncommon to encounter them while navigating around it. Central Island was still an active volcano and often emitted smoke, adding to the charm and wildness of the primordial landscape that characterized this famous lake.

We landed at the Leakey research camp airstrip in Ileret. We would be guests of Louise, Richard Leakey's daughter, and her husband, Emanuel. Maeve, Louise's mother, a fascinating woman with incredible knowledge of the prehistoric area where we were, would also be at the camp. It was a great privilege for us to learn from these strong, passionate women about the earliest archaeological finds and the first hominids discovered in Kenya. Richard Leakey was a world-renowned anthropologist. His passion, vast knowledge, and perseverance had led to the most important discoveries in the world in terms of hominids and fossils.

The camp was in a very isolated and remote area in the far north of Kenya, near the Ethiopian border. Two shipping containers, hand-painted by Louise, with windows carved into them, formed the heart of the camp. They served as a central area for meals, an office, and a kitchen equipped with everything necessary to be functional and perfectly clean. We slept in canvas tents. Simply and sparsely furnished, they were comfortable and spacious, with bathrooms that had running water and electricity generated by wind turbines. The wind, hot and dry, which is characteristic of the entire Lake Turkana area, reigned supreme. It blew incessantly, day and night, lifting dust and dry shrubs.

After lunch, we decided to take a walk on the dry bed of an ancient lake where, if you looked closely, you could spot hundreds of fossils. It was an almost lunar landscape. The vegetation was sparse, desert-like, except for the beautiful desert roses in bloom. The dark, volcanic, dusty, and arid terrain was scattered with fossils rolling at our feet. Walking slowly, we saw bones of prehistoric animals that had lived in that very place millions of years ago. Holding some in my hand, stroking them, and listening attentively to Louise and her mother's explanations, I was filled with a deep peace and a sense of respect and mystery.

Only now did I truly understand that life as we know it had originated here. Right on this piece of land, beneath this infinite blue sky, with the howling African wind, the first hominids had come to life – the very beings from which all of humanity had descended. Maeve had recently discovered, not far from our camp, the skull of a hominid dating back 3.5 million years. This was the "Place Where It All Began," and we were thrilled to have come here.

After dinner, we sat for a long time, savoring the beauty of the African sky. Nowhere else in the world had I seen such an infinite number of stars; their brightness, their twinkling, and their glimmering left us speechless, immersed in the mystery of nature and the universe.

The following afternoon, at sunset, we went for a walk along the lake's shore. We strolled hand in hand on the fine sand, which smelled of soda and fish. The sky, painted in reds and oranges, was alive with pelicans and other water birds. We even spotted a pair of regal fish eagles calling to each other with their unmistakable cries. The Leakeys had taken a dip in the lake's warm waters. Luca, Jake, and Aisha followed them. I stayed on the shore, watching the baby crocodiles dive into the calm waters near me, playing in the waves, unsure of what I wanted to do... Unfortunately, my fear of unwanted encounters outweighed my sense of adventure, and I decided not to go in. After the swim, an ice-cold beer in front of the lake's tranquility and mystery tasted especially good! We waited for the sun to set in the lake before heading back to camp, retracing the tracks left by our Land Rover on the white sand, in the silence and magic of that moment.

The following morning, at dawn, after thanking our gracious hosts, we took off for the Chalbi Desert, where we would spend a couple of nights.

In one of the largest permanent oases in that desert, not far from the Ethiopian border, the local Gabbra people had built a simple yet charming lodge to host tourists eager for adventure and exploration of the surrounding desert. The Chalbi Desert was vast and inhospitable. It was traversed on foot or atop tireless camels by the fascinating and wild nomadic tribes of northern Kenya: Samburu, Rendille, Borana, Gabbra, and Turkana.

Kalacha Camp was constructed using only local materials, such as palm trunks for support structures and palm leaves for walls and roofs. It was managed exclusively by Gabbra women, who were kind, hospitable, and proud to showcase their land and culture to those interested.

The lodge stood on the edge of a spring. Thus, every morning at dawn, herds of camels would come to drink in front of our cottages. They arrived slowly, with desert dust ingrained in their light-colored fur and wise, weary eyes, knowing that finally, after several days of walking, they could drink and rest. They knew they were approaching water: their pace quickened as they neared the spring. They trotted happily over the last hundred meters separating them from the oasis, using their last energies, so preciously saved during their days of travel. Their owners, Gabbra or other nomadic people, wore colorful turbans covering their sun and sand baked faces, and stood nearby, silently watching them, never losing sight of their companions. It was clear they had a special bond with these animals. They knew them well, trusted them, and shared with them the hard and difficult life that fate had allotted them.

Immediately after, hundreds of desert partridges arrived from all directions. They glided over us, adorning the pink dawn sky with small dark dots, finally settling on the opposite side of the camels, drinking, bathing joyfully, and playing in the fresh water.

It was a spectacle that brought great peace and tranquility to my heart.

After lunch, we decided to take a walk to explore the nearby Gabbra village and visit a small Christian church built by missionaries many years earlier, where beautiful biblical paintings could be admired.

The next morning, we relaxed at the lodge. It was very hot, but fortunately, there was a small pool where we could cool off in the shade of large palms. In the afternoon, after borrowing a car from the missionaries, we went for an aperitif in the desert.

Never before had I seen the desert, and I was won over by its primordial charm. We stopped near a high, flat rock, and upon reaching the summit, we stood still, admiring the vast spaces illuminated by a warm golden light, with the sun setting before us, in perfect and absolute silence. We stopped talking – our souls were one with Mother Nature. And with the universe. I absorbed the magnificent energy of that ancient, mysterious place, as Luca embraced me. There was no need for words to describe that moment, only our hearts beating close together.

The day after, it was time to set off for a new destination on our adventurous aerial safari in northern Kenya: Samburu National Park.

After just over an hour of flying, we landed and were welcomed by an Italian friend of ours, Riccardo, who owned two lodges in Kenya: in the Maasai Mara and Samburu.

Riccardo's passion for Africa had begun in a rather curious way. He was born in Brescia, like Luca and me. As a child, he lived in the same apartment where Luca had actually once lived previously. Pure coincidence? Perhaps.

A talented journalist and writer, Riccardo had often visited Campi ya Kanzi over the years. He had mentioned to us several times his love for Kenya and his desire to move to Africa and build a lodge. He often repeated that we had inspired him – the choices we'd made and the life we lived. Riccardo finally moved to Kenya about 20 years ago; with tenacity, intelligence, and determination, he built not one but two beautiful lodges that have been highly successful.

The lodge where we decided to spend the night was built on top of a granite rock, with an incredible view of the arid, semi-desert plains of Samburu. It consisted of six cottages, made of sandstone and canvas. They were built in a very innovative and natural way: they blended in seamlessly with the surrounding environment and were superbly decorated with furnishings, antique brass lamps, and Moroccan rugs mixed with colonial and African pieces. It all came together harmoniously – welcoming, relaxed, and free.

Upon our arrival, we went straight on safari in Samburu National Park and stopped for lunch by a river populated by elephants, hippos, many birds, and monkeys. Riccardo prepared a superb picnic that we enjoyed in the quiet of the African heat, in the shade of enormous tamarind trees. That afternoon, we were lucky: a pair of young male lions came to drink in the river and then fell blissfully asleep on the opposite banks, right in front of us!

The next morning, while our men went for a walk, Aisha and I stayed at the lodge by the pool to relax and chat. The pool was a true work of art: carved into the rock, jutting out over the plains below. Swimming while enjoying the superb view of Samburu's endless plains was a great thrill. After lunch, it was already time to go home: time had flown by, and I was sorry to think that the vacation was over. But I couldn't wait to return to Campi ya Kanzi and hug my little ones. After a couple of hours of flying, we finally landed in the Chyulus. The kids had come to the airstrip to welcome us: they were excited and happy to see me again!

It had been wonderful to go on safari with Jake and Aisha. We had lived magical moments of total relaxation and unity with nature, and we, too, couldn't resist the charm of the North, the history of the first inhabitants of the Earth, the infinite arid spaces of the desert with its precious springs, and its brave and solitary people. For this, we thanked our friends for encouraging us to join them on another unforgettable adventure.

Nonetheless, being back home at Campi ya Kanzi, in our beloved Chyulu Hills, with our children and the Maasai, was the best thing of all. More and more, I felt connected to Africa and its people.

Sorcerers and Fortune-tellers

Once, someone read my palm and told me that around the age of 40, my life would undergo a tremendous change. They predicted that I would face a great choice in love, one that would require me to sacrifice something I deemed irreplaceable. Yet, they also foretold that this sacrifice would bring joy and light into my life, and that I would never regret it.

I had never truly believed in fortune-tellers, magicians, gypsies, or those who claimed to read the signs that the rest of us cannot see. Actually, I had always kept my distance from these types of people, perhaps because deep down, I feared what they might reveal and knew that, against my better judgment, they would influence me.

But the ones who told me of the great change that would come with a profound choice in love were not magicians or fortune-tellers; they were two close friends of mine who did not know each other at all, nor did they read palms for a living. They were, let's say, a bit like my grandmother Nella. They had a different kind of sensitivity, one that set them apart from others. This unique insight gave them the ability to peek into other's souls and see things invisible to the rest of us, things that would inevitably eventually unfold. And they both told me the same thing.

What struck me was that they both said this to me when I was 38 years old, for entirely different reasons and in completely different circumstances. Their words stayed with me, tucked away in my mind, and as I turned 40, I began to wonder if and what this choice of love would be.

I celebrated my 40th birthday, reflecting on myself: my life at that moment was full and happy. I had two beautiful and healthy children who brought me great joy. Lucrezia was growing up quickly; she was already seven, almost a young lady. Sweet, beautiful, sensitive, cheerful, and far more mature than her age. Jacopo, on the other hand, was a whirlwind. A blonde hurricane with lively green eyes. At just a year and a half, he was already a lightning bolt: utterly charming and always on the move. He explored everything, climbed everywhere, and sought out the most dangerous adventures to live!

Things were going well at Campi ya Kanzi, and our foundation was also seeing excellent results in various projects aimed at safeguarding the Maasai people and the flora and fauna of the reserve.

We had created the foundation - "The Maasai Wilderness Conservation Trust" - as a non-profit organization to help the Maasai community of the ranch where Campi ya Kanzi stood. Its purpose was to preserve the environment, the wildlife, and the Maasai culture. Funded by us and our families, our guests, and by international donations – through the foundation, we helped the Maasai community, particularly the schools and dispensaries in our area.

I had returned to working full-time at the lodge with our guests; I managed my boutique, Maasai Chic, the school projects of Kanzi Academy (the little school attended by my children and their Maasai friends), and the dispensary funded by the foundation. And finally – but not because they were any less important or demanding of my time – I devoted myself to my children and husband.

If someone had asked me how I felt during that period of my life, I would have said, "Content, fulfilled, happy, and satisfied!" Given the circumstances, I had no desire for any sudden changes or to make any strange choices in love.

But I had not considered the unpredictability of life, which is one of its most fascinating aspects.

In September 2008, I discovered that I was pregnant again. Needless to say, this pregnancy was neither sought out nor planned. It was completely unexpected, and it turned the calm and predictable course of my life completely upside down.

Never had I thought I would have three children. Above all, I never imagined having another child at 40, when my youngest was just over a year old and still needing my care and attention. My first reaction was one of total disbelief and shock. Then, I must admit, came the fear that I might not be able to raise a third child adequately, followed by the dread of starting all over again. The worry of facing another pregnancy, another birth, and another period of breastfeeding came next, with all the sacrifices and physical and psychological limitations I knew it would entail. I felt paralyzed.

The first few days after discovering I was pregnant were tough for me. I battled against my maternal instinct and the force of life that was already, firmly, making its presence known within me. Luca patiently stayed by my side. I could tell he would be overjoyed to have a third child. He only said these words to me: "Imagine how wonderful it would be to have another little one, curly blonde hair and blue eyes, happily pattering around the house". I must admit that this thought sparked something within me, leading me to finally embrace my new pregnancy with joy and love. And the most incredible thing was that my third child, Lorenzo, indeed had blonde curls and blue eyes! How Luca had known this before he was born, I still cannot explain...

A few days later, I remembered what my friends had said, reading my palm. "Around 40, a great change in your life, a great choice in love, will bring sacrifices but also immense joy..." My child, my third child: There he was, my choice in love! Now it was all clear to me.

After taking all of this in, I accepted my unexpected pregnancy with delight, never looking back or regretting the path I had chosen. The pregnancy went smoothly. Perhaps because it was the third or because I was so busy with my other two children, the day Lorenzo was born arrived almost without me realizing I was about to become a mother for the third time.

He too was born in the Nairobi Hospital, with the same doctors and nurses who had helped with my earlier deliveries. Luca was again by my side, helping me through the labor. I expected, precisely because it was the third, that this birth would be easier and quicker than the previous ones. I was wrong. It was the hardest and most dangerous. Lorenzo had the umbilical cord wrapped around his neck, and despite my pushing, he would not come out. They considered an emergency C-section, but I decided to give myself a few more minutes and a few more pushes to see if the baby could be born naturally, as had been the case with his brother and sister.

Fortunately, I was right. After long, exhausting hours of labor, Lorenzo came into the world naturally on May 12, 2009, at 8:15 in the morning. He weighed 3.7 kilos, was 52cm long, and was in perfect health.

As always, my mother and mother-in-law were in Nairobi to be with me and support me, and they flew back to the camp with me and the newborn a week after his birth. They stayed for several days until I felt more rested and ready to embrace my new role as a mother again.

Lorenzo was beautiful and calm, but always hungry. He nursed eagerly from my breast, every two hours, day and night, just like his brother. This was exhausting for me, especially at the age of 40! Aside from that, he was a sweet and gentle baby, who loved being in my arms and constantly craved cuddles.

Katito was fortunately by my side, helping me as she had with Lucrezia and Jacopo. Lorenzo was welcomed at five days old by our Maasai community with joy and love. They gave him precious gifts and bestowed upon him a Maasai name: *Mbatian*. It is the name of the greatest leader in Maasai history. I must admit that it suits Lorenzo perfectly, as now, at just 15 years old, he already shows leadership qualities, independence, and fearlessness.

Lucrezia immediately felt like a little mother to her youngest brother. She cared for him and helped me look after him with incredible tenderness and maternal instinct. Jacopo was surprised by his brother's arrival. I believe he was still too young to understand what was happening. In the early months, he was very protective and not at all jealous of Lorenzo. Sizing up anyone who came near his brother was his new favorite activity! He called him *"'Encio mio!"* and smothered him with hugs and kisses on the nose.

When Lorenzo began to walk and wanted to play with Jacopo's toys, the two of them started to bicker, scream, fight, and stake out their respective territory like two little lion cubs. After months of furious battles and brawls, they settled down and learned to play together and respect each other.

As I write, Lorenzo has just turned 16. As a child, he bore a striking resemblance to Lucrezia at the same age: very blonde, with curls, a perfect little nose, and huge blue eyes. He looked like a little angel from a Botticelli painting. Despite his angelic appearance, he has always been stubborn and strong-willed. Even as a child, he feared nothing and imitated his much older siblings' behavior without hesitation. He could also be incredibly sweet and loved being in our company, but he is probably the most independent and daring of my children.

It is fascinating to see how different my three children are from each other, how strong their personalities already are, and how they reflect remarkable similarities to me or Luca. Watching them grow, face life, and discover the world for the first time through their pure and sincere eyes, is captivating.

In hindsight, I believe three is the perfect number. Since Lorenzo arrived, there has been a new balance among my children and within our family. I have realized that Lorenzo was my "great choice in love," and I already see and experience, day by day, the immense joy and fulfillment he brings us. His birth was a great life lesson for all of us: always choose love, no matter the cost, even when it seems like madness and may involve great sacrifices. It is what gives life meaning. It is what will remain when everything else fades away.

Pongo and Lucrezia

He had arrived at Campi ya Kanzi on a windy September morning, moving slowly, with his head bowed and weary eyes, silently stirring up dust and the few remaining dry blades of grass. He was thirsty and hungry. He was a young bushbuck, and we named him Pongo (the Kiswahili name for his species).

Lucrezia had instinctively rushed to the kitchen and grabbed a leaf of lettuce, stretching out her small hand towards him. Pongo leaped back, startled and frightened, retreating a few meters before cautiously approaching her again, slow and wary. There was an immediate bond between them. Lucrezia remained calm in his presence, showing no fear of being hurt by the wild animal. We were struck by how at ease and relaxed our daughter was, even so close to a creature of the wild. I felt a surge of pride and watched with admiration as she tried to befriend Pongo. She never doubted for a moment that she would succeed, and that she and Pongo would become inseparable.

Pongo took time to trust my daughter. But after a few days, he grew accustomed to us. He always arrived at the same time in the late afternoon, slowly approaching the kitchen and patiently waiting outside for his food. Lucrezia was the only one he trusted, the only one he let near him. After about a week, he was eating lettuce leaves from her hand without alarm, without fleeing. After a month, she even managed to gently stroke his damp nose, and she was deeply moved. They had become good friends, trusting each other and unafraid of their differences.

It was late September 2009, the year of the great drought. There was no rain in April and May, leaving the land parched and desert-like under the relentless sun. We saw many animals dying of thirst and lack of grass and mourned for them. We bought vast amounts of hay from Nairobi, which arrived weekly with a rickety old truck. Every morning, we put out piles of hay around the camp for the animals. We made sure there was always plenty of water in the pond in front of the tents to quench the thirst of the creatures that regularly visited. Yet, the animals suffered, especially the buffalo and elephants, who needed vast amounts of grass and water to survive. We found carcasses scattered everywhere – zebras, antelopes, buffaloes, even hippos had vanished from the springs of Iltilal. It was heartbreaking and depressing to be surrounded by so much death.

Time seemed to stand still, and in its utter immobility, nature waited hopefully for a sign from the sky – a dark cloud or just a few drops of rain to announce the renewal of life, to let us hope and smile again. I prayed every day for the rains to come. I watched the sky and looked around: fine dust covered everything. There wasn't a single blade of grass left. Only the acacias held on to a few green leaves, waiting, enduring a little longer before dying completely. But we did not know how much longer they could last.

One afternoon, to our great surprise, we spotted a family of elephants, with a little one in tow, passing slowly in front of our house. It was the first time elephants had crossed the lava flows and ventured so close to the camp, so close to us. They were magnificent, majestic, and wise, struggling forward, heading from the dusty plains towards the Chyulu Hills, hoping to find water for the weaker members of the family, those at risk of dying.

The Chyulu Hills, usually green and lush, were now barren and dry, with little grass to be found. But the thick cloud forest on their peaks still offered some respite and a few green shrubs or leaves for the bravest, most determined animals that refused to die.

We watched the elephants, with their courage and incredible survival spirit, climb the hills to fight their battle against death. The little one walked slowly, struggling to keep up with the rest of the herd, but its mother stayed close behind, ready to support it if it stumbled. With deep sadness in my heart, I knew the little one would have not make it. I knew that if it died, its mother would stay by its side for days, trying to awaken it, gently caressing it with her trunk. But if she wanted to survive, she would eventually have to abandon it and rejoin the herd.

I had always had a soft spot for elephants, which were so similar to us humans in their behavior and herd dynamics. Their sense of family and the protection they always showed towards the smallest and weakest moved me deeply. They were my favorite animals, and knowing we would lose these, and many others, if it didn't rain soon, made my heart heavy and dark.

Once, we had saved a baby elephant: Chaimu. Chaimu is the name of the most recent lava flow in the Chyulu Hills, and it was here that the young elephant had been found. Our rangers had followed her tracks for a long time; they noticed there were none of her mother's. They realized she was an orphan, young, and in need of rescuing. For three days, they followed the droppings and tracks of Chaimu, but without success. The tracks disappeared into the lava, making it impossible to find her.

It was then that we asked an old Maasai for help. The old man gathered a handful of dust from the little elephant's footprint and wrapped it in his *shuka*, as if trying to capture the young one. With the dust enveloped in the *shuka* held tightly in his hands, he closed his eyes for a moment; then began to walk determinedly in the lava towards an unknown point. After a few hours of walking, he found Chaimu. He then unwrapped his *shuka* and let the dust from the animal's footprint slowly fall through his thin, wrinkled fingers. He didn't need it anymore; she had finally been found!

That is how the Maasai found lost cows. When we found her, Chaimu was in danger, thin, suffering, and blind in one eye. With the timely and invaluable help of the veterinarian, our rangers gave her milk and transported her by plane to the Elephant Orphanage in Nairobi, where she would receive the best care.

The day Chaimu was found, Luca took Lucrezia to witness her rescue. My daughter returned from that adventure with her heart full of joy and compassion.

In the following months, we often visited Chaimu to see how she was doing. The young elephant had recovered well; her eye had healed, and she would soon be ready to be reintroduced into the bush and live as a wild elephant again. This filled our hearts with joy. It made us feel like part of nature and gave us the awareness of how much we could still do to preserve our beloved Africa.

Perhaps because of what Chaimu meant to me, I had a deep bond with elephants. That is why, on that dry day, seeing the little elephant struggling outside my house, headed towards the Chyulu Hills, so weak and suffering, I was deeply sad.

Finally, that evening, the sky grew dark: large, black clouds heavy with rain gathered around the Chyulu Hills, surrounding them ominously. In Africa, the scent of rain is the first sign that the heavens will soon open up. The air fills with electricity, and the smell of water and life spreads everywhere. It is a sensation that immediately brings immense joy and euphoria to all the inhabitants of this mysterious continent, be they humans or animals.

That evening, the sky began to rumble, and in the distance, towards the Tsavo plains, lightning and thunder illuminated the parched earth as bright as daylight. The wind, blowing from the hills, began to howl, singing its song of rain at the top of its lungs. Heavy, thick drops of water began to fall on us, and the Earth was reborn in an instant.

It was the beginning of the rains: huge torrents of water poured over the camp. From the windows of our house, we watched the droplets for a long time, listening to their poetry of life renewed. The children began to dance and sing in the rain, and Luca and I embraced each other, happy.

Life had finally returned.

After a few days, the earth was covered with green, tender grass. Wildflowers, like white and pink cyclamens, bloomed, covering it with color and cheer. The animals returned to graze near the camp, peaceful and serene.

Pongo no longer came to the camp. Now that the plains had

plenty of grass and water, he no longer needed the food his friend Lucrezia gave him. My daughter, to my great surprise, never asked about him again – where he had gone or why he hadn't returned. I am sure she missed the little antelope greatly.

Lucrezia, in her deep wisdom as a daughter of Africa, understood that Pongo belonged to another world, and now that the Earth had come back to life, he no longer needed her. She accepted it without rebelling or mourning his loss.

Our Pongo never returned. I often catch Lucrezia silently looking in the direction her friend used to come from. I suspect she sends him a greeting from her brave and generous heart. I know she will never forget him. Nor will we.

Years have passed, and sometimes, at sunset, we spot an adult male bushbuck with long horns, standing still at the edge of the Chyulu forest. He seems to be watching us, silently. I like to think it is our Pongo, coming to steal one last glance of us and his friend Lucrezia.

Snakes and Scorpions

During my years in Africa, I was fortunate not to have encountered many snakes. They fascinated me, yet a deep, uncontrollable fear stirred in me when I saw one. Africa, I learned, was unforgiving to those with fears and phobias. It seemed that the moment someone confessed to a terror, the very object of that fear would begin to manifest around them. As if, instead of repelling what they dreaded, their fear called it forth as if Africa itself sought to test those burdened by such anxieties.

I remember one guest at Campi ya Kanzi who, upon arrival, declared her phobia of birds and feathers. I could not help but think she had chosen the wrong place for a vacation, as Campi ya Kanzi boasted more than 300 species of birds, a true paradise for birdwatchers. She settled into Tembo House, and soon after, a swarm of Superb Starlings – beautiful birds with iridescent blue plumage, like a starry sky along the coast – began circling the house in a frenzied dance as if disturbed by an unseen force. Their flight, normally graceful, became erratic as if they sensed her fear and wished to convey their disapproval.

Another time, an American guest confided to me that she had a fear of snakes. No sooner had she arrived at Tembo House than she rushed to the bathroom, only to emerge a minute later, white with terror. A puff adder—a thick, dark, and lethally venomous snake—was coiled in the corner, shrouded in darkness. I marveled at her luck in spotting it before stepping on it, for these vipers, which are sluggish yet surprisingly agile, are responsible for more human fatalities in Africa than any other snake. They often strike as a defense mechanism, usually because someone has unknowingly ventured too close or stepped on them in the dark. I greatly admired her composure in that moment of terror.

To our astonishment, when we escorted her to her tent, we discovered another snake beneath her bed. Never, before or since, have we encountered snakes in the camp, let alone inside the tents. It seemed surreal, as if the very presence of someone so frightened by snakes had drawn them out from the shadows, like phantoms summoned by terror itself.

But one day something strange happened to me. I was eight months pregnant with Lucrezia and, for reasons I could not explain, I had been dreaming of snakes, feeling a constant, subtle dread. One afternoon, as I rested in my tent, I noticed a small snake slithering towards me. My heart froze for a moment, but somehow, I did not panic. I stared at it, meeting its gaze, unmoving for a minute that felt like an eternity. Then, slowly, I rose, never breaking eye contact, and opened the tent door. I thought, "This is not the place for you... go outside." The snake hesitated, then slithered cautiously towards the door, still watching me, before disappearing into the outside world. The memory of that encounter, my first close call with a snake, is etched in my mind, as vivid now as it was then.

There was also an incident with an Italian travel agent, during the early days of Campi ya Kanzi. We often hosted trips to introduce the camp to tour operators and agents, hoping to attract clients and grow our business. On that particular morning, I was alone at the camp; Luca was in Nairobi. I was eagerly awaiting the arrival of some important Italian travel agents. I wanted everything to be perfect and to make a good impression because we needed the business.

They finally arrived. I greeted them warmly and invited them to Tembo House for a drink. One of them, the most important, approached me and asked, "What's the snake situation like here? My clients are terrified of snakes, and it's crucial that the places we send them have none." Naturally, I assured him, "Don't worry, we don't have any at all." As I spoke, out of the corner of my eye, I saw a small, non-venomous house snake drop from the roof behind him. He didn't notice, but I did!

Once again, I marveled at my own calm and quick thinking. I casually directed his attention to Kilimanjaro, which stood majestically before us, while, with a swift kick, I discreetly shifted the stunned snake away from the veranda. I continued chatting amiably, sharing amusing anecdotes about Africa and the Maasai, while making sure the agent moved towards the terrace, away from the snake. I quickly called Patrick, our cook and the resident snake handler. Patrick arrived swiftly and silently, capturing the snake outside Tembo House without anyone noticing. I must admit that I felt a pang of guilt for my little lie, but under the circumstances, I had no other choice. The agent never knew what had happened and went on to send us many clients after that visit.

I may have been lucky with snakes, but scorpions were another matter. I was stung by a scorpion only once, but it remains the most painful and traumatic experience of my life, more so than even three natural childbirths without anesthesia!

It happened in early April 2010. The rains had begun, and the world was turning green and blooming. As usual, at the start of the rainy season, snakes, insects, and scorpions emerged from months of dormancy.

Lorenzo was almost a year old. I had stopped nursing him, and he no longer slept with me. He had his own room and a nanny who patiently watched over him during his long, restless nights. I was grateful for the help of my dear Maasai nannies, such exceptional caretakers for my little ones. Lorenzo, like Jacopo at his age, did not like to sleep. He often woke up, in the middle of the night, in tears, haunted by unknown nightmares or unseen discomforts. Since having children, I had become a light sleeper, and at the sound of his cries, I would rush to his room to make sure everything was all right and to soothe him back to sleep.

That night, Lorenzo's cries woke me as usual. Still groggy, I headed barefoot to his room, carrying a flashlight just in case. After calming him down and making sure he fell back asleep, I made my way back to my bedroom. Suddenly, I felt a searing pain in my toe, like an electric shock, that stopped me in my tracks. In agony, bent over and on my knees, I turned on the flashlight to see what had stung me, but saw nothing.

Hobbling back to my room, I woke Luca and told him what had happened. At first, I feared it might have been a venomous snake. Luca inspected my foot but found no bite marks. Meanwhile, the pain intensified monstrously. I could feel the venom coursing up my leg, through my femoral artery; the glands near my groin throbbed. My heart raced, cold sweat drenched me, and I trembled uncontrollably. I lost all feeling in my foot and began to spasm, crying out in agony. Never had I experienced such intense, unrelenting pain. From the stories of friends who had been stung, we realized, even without seeing the scorpion, that this was what had happened.

We had nothing at home to treat scorpion stings. It was the middle of the night, and we could not fly to the hospital in Nairobi until morning. It was going to be a long night. I remembered someone once telling me that soaking the foot in boiling water could alleviate the pain. So, with great difficulty, I was carried to the bathroom and placed my throbbing foot in a basin of boiling water. Fortunately, it helped a little. The pain became slightly more bearable.

Word quickly spread around the camp, and by 2 a.m., a cluster of Maasai arrived from the village, asking to see me. They assured me they knew how to ease my pain. I let them into the bathroom, and they said the only remedy was to spit on the wound. However, the saliva had to be special, not just any Maasai's. They had brought the village's scorpion bite specialist, whose saliva was considered the best for the job. With his help, they said, I could be certain of feeling much better! I must admit, I was briefly tempted to let them spit on my foot. But then, reason took hold. I thanked them for their thoughtfulness but returned to the remedy of my boiling water even if in relentless pain.

The next morning, after eight or nine hours of keeping my foot submerged in steaming water, my skin began to show signs of burns. Luca had tried in every way to convince me to take my foot out of the basin, but I was adamant to keep it submerged. The only way I could dull the pain was by keeping my foot in the scalding water. I was beyond reason; I was angry, suffering, and aggressive. Exhausted from a sleepless night and devastated by the merciless pain, I was nearing my breaking point.

Then Katito, my African sister, arrived. She sat beside me, gently stroking my head and speaking to me softly, as one would to a child. Somehow, she persuaded me to remove my foot from the basin.

What followed were endless hours of sharp, stabbing pain. I succumbed to it, slipping into a world of cold, dark shadows. Our pilot had flown to Nairobi to purchase medication, but she returned only in the evening, after I had endured the brunt of the pain and was absolutely spent.

Katito had told me that the pain from a scorpion sting lasts 24 hours. Surprisingly, she was right. Just past midnight the next day, I began to feel better, much to my relief. However, I could not walk for over a week. The pain in my little toe was intense for many days, and I lost sensation in my entire foot for a long time. I felt poisoned by the scorpion's venom. It was inside me, not just in my foot, but everywhere. I was unwell for days. But I survived, and after the scorpion sting, I felt even more like a daughter of Africa!

Unfortunately, my daughter Lucrezia also experienced the pain of a scorpion sting. But she was luckier and much braver than her mother. It was a beautiful December morning, and Lucrezia had woken up happy and excited. She was about to fly to Nairobi with me to finally meet her new pony named Piper.

My daughter had always loved horses ever since she was very little. Neither Luca nor I shared this passion, but that was how it was: Lucrezia never missed an opportunity to ride. The walls of her bedroom were covered with posters of magnificent horses, and she had more than 50 pieces in her collection of toy horses and "My Little Ponies".

That year, she had done very well in school and was at the top of her class. So, Luca decided to reward her by giving her what she wanted most: a pony of her very own.

After searching for the perfect friend for Lucrezia, we chose a beautiful pony that was brown and black. We bought it, and that morning we were heading to Nairobi to finally show it to Lucrezia. While getting dressed, my daughter screamed and quickly pulled off a riding boot. She cried, "Mom, mom, there's a scorpion in my boot!". I looked inside the shoe, turned it upside down, tapped, and shook it, but nothing came out. So, I reassured her and urged her to put it back on. We were late, and we needed to go. As she put the boot on again, Lucrezia screamed once more and began to cry uncontrollably, "Something bit me, mom. I'm sure it's a scorpion!"

Unfortunately, she was right. When I turned the boot upside down again, a scorpion finally fell out. Fortunately, my daughter was wearing thick socks, and the scorpion probably couldn't inject its venom deeply. However, Lucrezia was crying and screaming. She was in a lot of pain, and only I knew how much. I was so upset to see her suffer so much. I felt powerless in the face of what was happening. I would have done anything to ease my little girl's pain.

This time, I had everything I needed. I immediately gave her a strong painkiller and applied an anesthetic ointment to her foot. After about half an hour, Lucrezia began to feel better. She calmed down and stopped crying. An hour later, she was back on her feet. She was limping, but she could walk. She certainly would not be able to wear shoes for a few days, but she was determined to meet Piper and become lifelong friends. And so, feeling proud of my daughter's strength and willpower, I set off with her to Nairobi.

After landing at Wilson, we headed to Karen, to the stable where Piper was kept. It was worth it: the immense joy and excitement I saw in Lucrezia's eyes when she met her pony made up for all the anxieties and worries of that morning. Once again, I had learned from my daughter to be brave and not to let difficulties get the best of me. Once again, Africa had taught me a life lesson.

Are you Married, or Do you Live in Kenya?

It is said that in Kenya, especially in Karen – a suburb of Nairobi named after the famed Baroness Karen Blixen, author of *Out of Africa,* who used to own a farm there – the KCs, the so-called Kenya Cowboys, and others know how to savor life, indulging in every form of pleasure. A sometimes affectionate, sometimes disparaging name, Kenya Cowboys are the boisterous, high-spirited descendants, male and female, of the original white settler community in Kenya.

From the early days, whispers spread across Kenya of marriages that were anything but conventional – open unions where both husbands and wives had numerous lovers and where forbidden romances and intoxicating champagne or whiskey soirées rivaled the golden days of Hemingway or Fitzgerald. Mysterious crimes of passion added to the allure. If you have seen *White Mischief*, you know what I mean! Those who left Europe to settle in the colonies in the early 1920s were eager to relish every moment of their new adventure in this exotic land, so full of inspirations and excitement that it would forever be etched in their hearts, even after they no longer lived there.

The atmosphere in those early days was so libertine, joyous, and excessive that upon meeting someone living in Kenya, one might ask, "Are you married, or do you live in Kenya?" This question hinted at the reality for many couples in the Happy Valley of Kenya. They married, had fun, took one or more lovers, but rarely divorced. Sometimes, however, someone who had taken their adulterous antics too far, or who had a particularly jealous spouse, would end up dead in some strange and unbelievable way. But no one was too surprised - it was all part of living in Kenya!

Today, the "typical day" for most residents of Karen involves working (though in moderation!) while still finding time to regularly play polo – many of those who live in Karen own horses – golf, or tennis at the famous Karen Golf and Country Club.

In the morning, residents shop at the Karen Provision Store, the first small supermarket in Karen, next to the gas station. Pyush is the area's most well-known hardware store and also a meeting place for many Karen residents. It is the go-to spot for buying tools for DIY projects, garden supplies, or even food and shelter for pets. Practically anything needed to keep one's home and garden in good condition and to make their dogs or cats happy, can be found at Pyush.

Tea served at 5 p.m. is a true tradition, a not-to-be-missed ritual for anyone living in Kenya. *Chai* (tea in Kiswahili) is usually enjoyed with milk and sugar (following English tradition) and is often accompanied by delicious pastries or tiny sandwiches filled with mayonnaise, cucumbers, and ham.

Instead of tea, one might stop by the Purdy Arms for a beer, if only for a few minutes. It is a chance to meet a friend, catch up on the latest news and Nairobi gossip, discuss politics, and, above all, relax.

Alternatively, one might go for lunch or dinner at The Talisman, a historic restaurant in the heart of Karen, with a welcoming atmosphere that is both cozy and sexy, surrounded by century-old pepper trees and colorful bougainvillea. The restaurant offers delightful international cuisine with an exotic touch, as well as art exhibitions by African artists, attracting many customers with an interest in art, or simply looking to enjoy a meal with friends in a different and captivating place.

In the early 1900s, it was said that incredible parties were held in Karen, in Muthaiga, and along the shores of Lake Naivasha, north of the city, in the large and beautiful homes of the first westerners to arrive in the country. Surrounded by lush tropical gardens and hidden from the view of the curious and outsiders, these parties were the reason behind the area's name - Happy Valley.

Muthaiga has always been the residential area where embassies are located and where numerous NGOs and the United Nations have their headquarters. It is considered the most snobbish area of Nairobi. The cost of land in Muthaiga is comparable to European prices. Beautiful houses, immersed in the tranquility of lush gardens, make Muthaiga an elite and exclusive neighborhood. The perfectly kept driveways provide access to the colonial-style villas located in this prestigious area. The famous Muthaiga Country Club, founded in 1913, is still in fashion and perfectly maintained, just as it was at the beginning of the century. Its membership selection is very strict, and it sometimes serves as a meeting place and retreat for wealthy and successful individuals.

We often stayed at the Muthaiga Club in our early African years, when we were building the camp and had to spend a few nights in Nairobi. Dressing up for dinner in the evening was a must. The men had to wear jackets and ties, and the ladies a cocktail dress. It was like reliving the atmosphere of the early 1900s, when figures like Karen and Bror Blixen, Dennis Finch Hatton, or Lord Delamere were regulars at the Club, making it irreplaceable and unique. I enjoyed going to the Club. It was an opportunity to finally savor a decent meal and enjoy a good glass of wine in the company of dear friends, making my early years in Africa less burdensome.

Karen, unlike Muthaiga, has always been the area near the Ngong Hills forest where the KCs predominantly lived. It was ideal for farmers who lived on the highlands and decided to spend some time in the city. The houses were mostly stone cottages, with corrugated iron or grass roofs, often smaller and less pretentious than in Muthaiga. The homes in Karen were undoubtedly less refined and expensive than those in Muthaiga. The people who lived there were simpler, less sophisticated. The century-old virgin forest at the foot of the Ngong Hills still reigned supreme, and the roads leading to the various houses were often dirt and poorly maintained. Occasionally, solitary leopards and mischievous monkeys would play in the gardens of those who lived in Karen. When night fell, those with dogs or cats made sure they slept indoors, safe from predators. The bush babies and hyraxes let out their mournful cries every night, and their wide, unsettling eyes gleamed among the branches of the enormous century-old trees in the nearby forest.

It was all wilder and less tamed than in Muthaiga; but maybe this is exactly why Karen was so enchanting.

This small suburb, just a few kilometers from the chaotic and modern center of Nairobi, could be considered a legendary spot and had subsequently attracted many famous figures, including artists, writers, intellectuals, and others who made their homes there. One could say that Karen had its own story, its own magic, its own soul. In Karen, one could pick up on an atmosphere that was both exciting and decadent at the same time.

Karen was the area of Nairobi, along with Muthaiga, where the first settlers established themselves at the beginning of the century, and where the country's social life took place, with its elegant and refined parties, always impeccably planned and sometimes excessive!

Although the parties in Karen today are no longer as dazzling, the spirit is the same as back then, and so is the desire to have fun and enjoy life. Today, the gardens of Karen's houses may be a little less maintained, the crystal glasses a bit chipped, the silverware dented, and it is rare to be invited to a party where gentlemen wear tuxedos, and the ladies wear long gowns. The inhabitants of Karen no longer move about on the backs of beautiful horses, but mostly in old and battered Land Rovers. The men wear comfortable *kikoi* or khaki Bermuda shorts; the women wear wide cotton skirts and comfortable flip-flops. Of course, this does not apply to everyone, but many in Karen, particularly the 'KCs' (Kenya Cowboys) still live like this, sometimes breathing in the nostalgia of a Kenya that no longer exists.

After about ten years of our arrival in Kenya, we decided to buy a house in Karen. This way, when I was tired of the solitude of the savannah, I could go to Nairobi and spend a few different days and feel "urban" again, even if only for a moment.

Living in Karen is like being on vacation all the time. There is a jovial, relaxed atmosphere everywhere. Almost everyone knows each other; we meet in the usual places, have children who attend the same schools, go to the same doctors, and share in the joys and sorrows of friends who belong to the small community living in the same area.

Alongside the modern shopping centers that have now sprung up all over Nairobi like mushrooms, the *dukas* (small stalls of vendors selling fresh food or other goods) on the sides of the roads are still present and remain the destination of many Kenyan shoppers today.

In Karen, next to the main roundabout, there are many of them. It's always a fascinating contrast for me to see the multitude of noisy, busy African people eager to close deals among the colorful stalls; the matatu drivers constantly honking their horns, the clouds of dust raised by the wind, the cries of children tied in a bundle on their mothers' backs, dressed in colorful cotton clothes with headscarves, and the modern shopping malls next to them! Yet, the two realities coexist side by side undisturbed, both vibrant and engaging.

I still find Nairobi to be a pleasant place. Despite the impossible traffic, the noise, the crowd of people surrounding you everywhere, the pollution, the broken and dilapidated roads that turn into rivers in the rainy season, the "green city in the sun," as it was called at the beginning of the century, is the stage for a unique and fascinating life. From the barefoot Kenyan pushing a cart with a donkey to the wealthy politician in an expensive SUV with a driver and the latest model smartphone, everything is in harmony (or rather, fits together like the pieces of a puzzle) in the absolute chaos that reigns supreme in the city.

New neighborhoods are cropping up in Nairobi, beyond Karen and Muthaiga, where beautiful restaurants and clubs can be found. The nightlife is vibrant, and there are ever more tall modern skyscrapers made of glass and steel standing out against the city's blue sky. These are the areas of Westlands and Parklands, where the most prestigious offices in Nairobi are located and where, in the evening, hot spots fill with people of all races and cultures, united by a shared desire for fun and entertainment.

Nairobi is an incredible, international, and cosmopolitan metropolis that grows in population every year. Its economy is continuously expanding and, consequently, offers significant job opportunities to those from around the world who can seize the prospects that this city has to offer.

Yet, the reputation of the Happy Valley (whether deserved or not) still lingers in the air. Glistening silver pieces and bubbly champagne still take center stage here!

Kiwayu: a Dream that Feels Like Reality

There are many magical places in Kenya, dear to our hearts, that stir within us profound and unique emotions. But none are as beautiful as Kiwayu. It had always been so, ever since we first arrived in Africa.

Whenever we felt the need to escape the routine of camp, to refresh our souls, to nourish our spirits, to rediscover ourselves, even if only for a moment, in that incredible "world of dreams," as when we were children, living in wonder and awe—our destination was always the same: Kiwayu.

The island of Kiwayu lies in northern Kenya, near the border with Somalia. In front of this island is a strip of unspoiled land, caressed by crystalline, warm waters and a gentle wind, fragrant with salt and the sea. A coral reef encircles and shields its splendid bay from the persistent waves and tides of the ocean.

Kiwayu's beauty is an astonishing spectacle of nature, never ceasing to amaze us, never failing to open our hearts. Inland stands dense and tangled tropical vegetation composed of palms and dangling vines, and acacia trees with thorny branches grow at the base of soft sand dunes. Along the coast, the sand glows blinding white under the midday sun, while the sea shifts endlessly between shades of turquoise, indigo, emerald, and pale blue, ever-changing with the light and the tides, a sight we could never tire of admiring.

The beach, soft and fine as talcum powder beneath our feet, is dotted with tall, proud palms standing like sentinels along the shore, tousled by the warm breeze blowing in from the Indian Ocean. Curious baboons and mischievous vervet monkeys are frequent visitors.

In the dense mangrove forests, calm, sleepy channels open, traversed by expert Bajuni fishermen, gliding in their fragile canoes like nutshells, searching for crabs and delicate shrimps. Pink, graceful fiddler crabs, the dancers of the sea, cautiously emerge from their burrows at dusk to revel in the light of yet another fiery sunset.

We'd take long walks along the promontory, revealed by the receding tides, where brilliant red starfish, corals, shells, seaweed, and driftwood—and sadly sometimes even plastic bottles and broken jars—were washed ashore on the pristine beach, reminders of life and death, of time passing, and a world in constant change.

There were often also baby sea turtles, newly hatched, hesitantly heading towards life, bravely diving into the warm waters of the ocean.

From the sea, we'd gaze upon the white sand dunes that cradled our makuti-thatched huts, where happy vervet monkeys leapt and played. The dunes, crowned by fragrant bushes of blinding green, seemed like precious lacework—delicate creations by wise grandmothers long gone but still very much alive within me, cradling my heart.

And the laughing African children, dressed in bright rags, riding their trotting donkeys along the fine, untouched sand, with ten-liter yellow plastic containers of Elianto oil strapped to the animals' backs, greeted us every morning with mischievous smiles as they travelled great distances from their small villages to the well to fetch water, proud to help their mothers, who waited on the shore, fishing for tiny, silver fish to cook for their families.

In the early mornings, the white triangular sails of *dhows*—traditional wooden boats, much like those of the ancient Egyptians—stood tall against the rosy dawn. The fishermen sang their age-old songs, repeating rituals unchanged yet ever new, hoping, as always, for a "good catch."

Then, there were the timid sunrises, savored from atop the sand dunes, trembling with the anticipation of a new day; and the flaming sunsets, as if molten gold had been poured onto the sky, witnessed from the veranda of our palm-thatched bungalow by the sea, as we listened to the waves crashing on the shore before us.

In the mystery of the nights, the full moon, so immense and white, would illuminate the ocean, bathing it in silver, like a path without end, leading to the edge of the world where sea and sky become one.

This, and much more, was Kiwayu to us.

Our dear friend Alfredo Pelizzoli, who had come to Kenya from Italy in the 1950s, discovered this slice of paradise, and we were eternally grateful to him for showing it to us. Alfredo had been a great professional hunter in Kenya, and it was on these white beaches of Kiwayu that he had set up his first tented camp for the hunters who, thrilled, came from distant Europe to capture trophies of exotic animals and take a piece of that incredible Africa back with them.

I remember him telling us of lions coming to drink in the moonlight on the beach, and of elephants swimming across the bay each evening to reach the island, safe from the dangers of man. What a paradise it must have been then...

After hunting was banned in Kenya in 1977, the rustic camp was transformed into a charming collection of bungalows for fortunate few tourists. Built of wood and palm leaves along the sea, the bungalows had large windows opening wide to the endless ocean, woven straw mats for floors, comfortable hammocks for lounging, and giant clam shells for basins to wash hands and feet.

At sunset, we delighted in sipping icy *dawas* (a Kenyan cocktail made with vodka, lime, honey, and ice), lounging on large cushions covered in vibrant kanga fabrics, leaning against old, sea-worn driftwood brought ashore by the tides. We watched the sun dive into the iridescent ocean, painting it with vibrant reds and shimmering yellows, and waited for the pale moon to rise, talking of trivial matters or asking existential questions—always so relaxed, always at peace with ourselves.

The décor of the bungalows was deliberately simple yet of great elegance and refinement. There was no need for many furnishings; everything necessary for a comfortable and unforgettable stay was already there. The infinite turquoise sea, the white sand, the blue sky, the warm breeze, the tall palms, and the green vines—all were the true protagonists in this scene, more than enough to give the bungalows a magical atmosphere. Everything felt so simple and natural as if those huts had stood on that beach in perfect harmony with it since the dawn of time.

What struck us most about Kiwayu was the overwhelming sense of freedom that filled the air. There was no need for clothes or shoes. A swimsuit (or even that wasn't necessary!) and a *kikoi* were all one needed. To remain barefoot, playing with the sand between our toes, feeling the warmth of the sun on our skin and the fragrance of the wind in our hair— was exhilarating. It felt like returning to the beginning of time. We felt free and happy, light and in harmony with nature.

I remember one particular night, the sky adorned with the brightest full moon. Its light woke me, so intense and dazzling that it cast shadows of palms on the white beach. The moon was enormous and perfectly round; if I stretched out my hand, it seemed I could touch it. It reflected off the calm, sleeping ocean, serene and silent. I rose and stepped outside, naked, walking across the sand until I reached the shore. The warm wind caressed my skin, and the moon, that enormous, magnificent moon, bathed me in its magical light. I spent a few moments inhaling the scent of salt, my eyes closed, in absolute silence, listening to the breathing of the sea and the waves breaking on the coral reef. Everything was so still, yet so alive...I felt so fortunate to be there, knowing I was experiencing one of those rare, perfect moments, so beautiful they seem like a dream from which one fears waking.

When we decided to bring our children to the sea for the first time, I had no doubt where we would go. Just a few days were enough for Lucrezia, Jacopo, and Lorenzo to fall in love with Kiwayu as deeply as we had. Jacopo took his first steps on that white sand and splashed, wide-eyed with awe and excitement, in the warm waters that lapped at Kiwayu's shores. Lorenzo played joyfully and carefree in the soft white sand, building castles with his father, populated by fairies and elves. Lucrezia gathered precious shells and driftwood along the shore to make necklaces and bracelets as gifts for me.

Kiwayu is our secret corner of paradise; it is our water when we are thirsty, our food when we are hungry. Kiwayu is the song of our souls, soaring high in the sky, like seagulls of distant seas.

Kiwayu is a dream that feels like reality.

African Friends

Over the years spent in Africa, we had the great fortune to meet many fascinating and kind-hearted souls who quickly became our African friends. Each one, in their own way, carried a story that made them special and unique in our eyes. But the dearest of friends, those closest to our hearts, were undoubtedly those who lived in Laikipia, one of the most beautiful places in northern Kenya.

The air on the highlands of Laikipia is fresh, light, and crisp. The altitude there is around 2,000 meters: the sun blazes during the day, but a gentle breeze from Mount Kenya refreshes the air, making the heat pleasant. The sky is clear, a blue so deep and perfect as I have only ever seen in Africa. The light and vivid colors that bathe Laikipia create a brilliance and brightness found nowhere else in Kenya. I have always wondered about the cause of this strange optical phenomenon. Is it the particular inclination of the sun's rays at the equator or the rarefied air that makes everything so magical and special?

Mount Kenya dominates the landscape, towering and majestic, with its broad slopes and sharp peaks, always dusted with a sprinkle of snow like icing sugar on a cake. Thick clouds, heavy with rain, encircle it like graceful ballerinas in white tutus, dancing and gathering around it at dawn and dusk. In the evening, when the sun goes to rest, the air on the highlands suddenly turns cold, and the sky ignites with a thousand bright, twinkling stars, so close it feels as if you could touch them with a finger! It is a sight that moves me every time, making me feel privileged to live in Africa.

Laikipia is a land of farms, and cattle and sheep ranches. It is the heartland of the KCs, the Kenya Cowboys. The first settlers who arrived in Kenya from Europe at the beginning of the twentieth century all lived in the northern part of the country, beneath Mount Kenya. The reason they stayed was for the soil: red, fertile, and rich in humus. Everything grew well and quickly; there was an abundance of rivers and rain, and the pasture was lush and green. In short, it was the ideal place to start a farm or raise animals. It was the perfect place to start a new life.

I always loved going to Laikipia. Like where we lived, Laikipia had been Maasai land, and perhaps for this reason, I felt even more connected to this piece of Africa. From the Chyulu Hills in southern Kenya it was about a two-hour flight. Luca had many friends up there, met during his first trips to Kenya. We tried to visit them as often as possible, though with our work, it was difficult to find the time. I always gladly accompanied Luca; I knew that seeing those dear friends and spending time with them would bring me happy moments.

There were Julian and Jane, who lived in Nanyuki, in a splendid house right under Mount Kenya. Julian had come from Scotland in the early 1950s. He had been one of the best professional hunters in Kenya. He had taken great figures hunting, including William Holden, Stewart Granger, and others. Those were the days when Hollywood stars competed to come to Africa for big game hunting and returned home with enormous trophies to show off to friends and numerous splendid lovers.

Julian and William had become good friends and, along with others, had created the Mount Kenya Safari Club. It was one of the most beautiful and famous lodges in all of Kenya. Even today, its fame precedes it. It sits near a nine-hole golf course and consists of a clubhouse and several guest cottages and villas. Built of wood and stone, all enjoy a magnificent view of Mount Kenya. The grandeur of 1950s colonial architecture is augmented by a timelessly elegant country club atmosphere. Each accommodation has a stone fireplace, and the fire is lit every evening while the guests are at dinner. At sunset, the temperature drops suddenly, and the warmth of the fire pleasantly heats and brightens the room, adding a touch of romance! The linens are of the finest quality, specially imported from

Europe; the marble bathrooms feature immense cast-iron bathtubs with wrought-iron feet, and the attention to detail makes this place truly unique and special. When we were visiting Julian and Jane, we often went to Mount Kenya Safari Club for dinner, and it always was a memorable experience.

Julian was also a friend of George Adamson. George was a great figure in the history of Kenya. He had accomplished countless feats in the conservation and protection of wildlife, especially lions. One day, he saved a lion cub, kept it with him, and, along with his wife Joy, raised and nurtured it until it was ready to be released back into the wild. It was a female, and they named her Elsa. Elsa's touching story was so incredible that it was made into a film entitled "Born Free." I remember how passionate I was about that film as a child!

When Elsa grew up, it was time for her to return to the wild to which she belonged. George and Joy took her to Meru, a large reserve in northern Kenya, where Elsa was released to live as a wild lion again in the African bush. However, George did not forget her: he often visited her to see how she was doing. He would land in Meru with his small plane and immediately fire a shot into the air with his old rifle, alerting Elsa to his arrival. Within half an hour, the lioness would appear, happy to see her old friend. Always.

One time, though, she did not come. So, George decided to spend the night in Meru in a small, tented camp and to leave his tent open in case Elsa would decide to come to see him later. A few hours later, at night, she arrived at George's camp, bringing along a "boyfriend". Sleepily, George felt a lion sniffing at his feet, and thinking it was his friend Elsa, reached out to stroke her. To his surprise, he felt a huge mane, one that Elsa definitely did not have! He realized with a fright that it was a wild lion that he had been getting accustomed to instead! Could it be that Elsa had been able to convey to her companion that George was a friend and not dangerous at all? George always claimed that lions could communicate with each other through their language, just as we humans do with words. Perhaps this was the proof...

Often, Julian accompanied his friend George to Meru. Each time, he watched in amazement at the miracle of friendship and affection between a wild lioness and a man. Over the years, George continued to spend most of his time with lions. He saved many orphaned cubs by nurturing and caring for them and releasing them back into the wild as soon as they were ready to hunt on their own. At the end of his life, those who knew him well, like Julian, described him as "more lion than man." For George, this would have been the highest compliment!

It was wonderful to listen to Julian's incredible stories about George's amazing friendship with his lions. Every Christmas, Julian and Jane would visit us at camp. In the evening, around the fire, under a sky sparkling with stars, Julian would begin his tales. We listened in silence, like children hearing the most beautiful fairy tales, overwhelmed by nostalgia. They were stories of the last glimpses of an Africa that no longer exists. Perhaps this is why we were so attached to Julian and Jane. They represented a dream that had vanished, but for this very reason, it was even dearer to us.

Julian was a discreet and elegant man, endowed with an incredible sense of humor. He still had great charm, even though he was no longer young, and he loved Africa and Africans wholly and absolutely. Especially the Maasai and Samburu. He had much in common with Luca and me: this was why, despite the great age difference, we tried to see Julian and Jane as often as possible.

Jane was born in London to a wealthy family. As a girl, she accompanied her father from England on a hunting safari in Kenya in the early 1960s. That is how she met Julian, who was a professional hunter at the time. They fell in love immediately. Jane left her country and settled in Nairobi for a few months, working at a travel agency. They married shortly after. Theirs was a simple but moving ceremony before the District Officer of Nanyuki and two witnesses.

Despite her age, Jane was still a beautiful woman: tall and graceful. She was strong-willed and determined, beneath an appearance of fragility. I had the impression that she knew well how to get what she wanted from her husband! She was always well-dressed and groomed and loved photography more than anything else. Her photographs, along with their company, were the best gift of every Christmas we spent together.

After hunting in Kenya was banned in the 1970s, Julian and Jane bought camels and began organizing safaris on the back of these animals in the north of the country. They travelled long distances on foot or by camel, reaching remote and still unexplored areas of Kenya. It was a life of adventure that required courage, patience, and infinite organizational ability. They catered to a select clientele eager to experience a Kenya still wild and far from tourist routes. They did it with passion for many years and had considerable success. Recently, they grew tired and decided to retire to a quiet life. So, they sold the company and camels with some sadness and regret and devoted themselves to traveling in Europe and their photography hobby.

Julian died recently, gored by a wildebeest while taking his usual sunset walk outside his home. They say the wildebeest, old and blind, perhaps crazed, charged him because it was frightened by Julian's unexpected presence nearby. I believe that, knowing our dear friend, this was the "perfect death" for a man like him, so in love with Africa, its nature, and its people. Jane has been left alone; they had no children. However, she is surrounded by the deep affection of her African friends, who care for and support her daily, making her heart, if possible, feel less tired and sad without Julian by her side.

Other dear friends who lived on the Laikipia highlands, in Timau to be exact, were the MacGregors. Graham MacGregor had met Luca in Naivasha back in '88. At the time, Luca was still searching for a ranch to buy in Kenya, dreaming of living in Africa. During summer vacations, or whenever he could, he would leave Italy to come to Kenya. He would sit outside the major banks in Nairobi, hoping to encounter some Kenya Cowboys, farmers or landowners dressed in khaki shorts and safari boots. As soon as he spotted one, he would introduce himself and ask if he could work on their farm in exchange for room and board. It was a chance for Luca to learn the farmer's trade, to discover that marvelous land, its people, and its terrain. It meant breathing in a bit of the Africa he so deeply loved. It also meant becoming a bit African himself, for that was, after all, his greatest desire!

One day, as he walked along a road towards Naivasha, a *matatu* (a minibus, used as the main mode of transportation in Kenya, meant to carry no more than ten passengers, though it often holds at least 20, all crammed together!) pulled over, and the African driver shouted at Luca in a contemptuous tone: "*Jambo Mzungu!*" Mzungu in Kiswahili means "one who runs around in circles without ever stopping." It is the term Africans use to refer to white people who come to Kenya. For them, accustomed to a life of calm and stillness, seeing people always in a rush, busy with a thousand things for no apparent reason, is a source of great amusement and laughter. Luca did not appreciate being addressed this way, especially the tone with which the driver spoke to him. He decided not to board the matatu and instead chose to walk all the way to Naivasha. It would have been a long walk!

Fortunately, only a few minutes later, two kind souls in a Range Rover stopped by the roadside. They introduced themselves: their names were Graham and James. Graham owned a farm in Timau, in the north of Kenya; James, his brother-in-law, was a lawyer and lived in Nairobi. They were on their way to play polo in Naivasha, a sport much loved in Kenya.

They gave Luca a ride, listened to his story, and understood his great passion and love for Africa. Graham then decided to offer Luca hospitality at his farm for a few months in exchange for his work. In truth, Graham not only opened his home, but also his heart and life to Luca with a spontaneity and immediacy that both surprised and captivated him.

It was the beginning of a great friendship, one that would bind them inseparably in the years to come. James, too, would become a dear friend of Luca's over the years, so much so that he would serve as the best man at our wedding.

Graham was not a man of many words; he loved to listen instead. His reserved nature hid a generous heart, full of affection and feelings, and an infinite kindness. He deeply loved his wife, Vera. Vera was strong. Tall, with straight blond hair bleached by the sun and blue eyes, she immediately gave the impression of being practical and energetic. Together, they had three children.

Graham and Vera worked hard together on the farm. They started by cultivating mainly wheat; thanks to Graham's intelligence, determination, and spirit of sacrifice, they had become one of the largest farms in all northern Kenya. They now exported all their products to Europe, mainly vegetables. They were well-known in Laikipia and admired and respected by all the other ranchers and farmers in the country.

When Luca first met Graham and Vera, Violet, their youngest daughter, was only a few years old. She was a little blonde girl with lively, clever blue eyes and fair skin, dotted with freckles on her little nose. She won Luca over instantly. While working for Graham, Luca enjoyed playing with his children during his free time, teaching them to ride a motorcycle and drive their Land Rover. He would hold Violet on his knee, telling her stories and teaching her a few Italian words, which she still remembers perfectly today!

Luca returned to the MacGregors every summer for many years, eventually becoming part of the family. Even today, so many years later, Graham and Vera remain dear and sincere friends to us, and we always enjoy visiting them. So much time has passed, yet to this day, I can see in Luca's eyes the respect and gratitude he holds for them, the sincere brotherly affection for their children, and a special tenderness for Violet, who has always had a special place in my husband's heart.

John and Olivia lived in Nanyuki, not far from Timau. Olivia was full of life and energy. Her eyes were bright and attentive; she loved to joke and often laughed. She was born in Botswana to a Scottish mother and a South African father. We met her when she was working for a tour operator in London and had come to see Campi ya Kanzi. She immediately fell in love with our lodge and sent many guests our way each year.

John was born and raised in Kenya and was a pilot. He obtained his commercial license in the United States. The son of friends of Graham and Vera's, he came to work at Campi ya Kanzi as a pilot. He was a very shy and quiet young man, honest and intelligent. It took time for him to open up, but after a few months of working with us, surrounded by the typically Italian sense of humor and Luca's and my desire to joke around, he became cheerful and witty. He was quite awkward with women and insisted he would never have a girlfriend, much less get married. He thought women brought nothing but trouble and worry!

One morning, we received an email from Olivia. She had decided to leave London and move to the United States, where her father now lived. She asked if she could spend six months at Campi ya Kanzi, helping us out at the camp before moving permanently to Boulder, in Colorado. We enthusiastically agreed.

Knowing her, always cheerful, lively, sparkling, and outgoing, I immediately thought she was the right woman for John. I told him so the very morning I received Olivia's email. "John," I said, "I've found the woman for you! She's coming soon. She's African like you: you're made for each other. You'll be crazy about her, and you'll marry her!" Naturally, his response was an incomprehensible muttering along the lines of, "Yeah, right, me getting married? Never! You're always quite the romantic!"

A few weeks after Olivia arrived at Campi ya Kanzi, they fell in love. None of us recognized the "gruff and solitary" John anymore, so taken was he now with Olivia, always attentive and kind to her. They were always together, laughing and joking constantly. He often brought her little gifts and fresh flowers from Nairobi; it was a joy to watch their love grow day by day.

After a couple of years, they got married. Happily. Today they live in Nanyuki and have a beautiful daughter. Unfortunately, we do not see them as often as we would like, but we keep in touch frequently and try to meet up as much as possible.

We also had some very dear friends, like mother and sisters to us, who lived in Nairobi and by Lake Naivasha, about an hour's drive north of the city. Emma, known as Picci to her friends, had moved to Kenya from Sicily in the 70s. Her parents had rented a farm on the shores of the beautiful Lake Naivasha and decided to start a new life in Africa. Picci, with three young daughters, had joined them in Kenya and settled on the lake with them, helping them with grit and determination in the cultivation and export of fruit and vegetables. Picci was a beautiful woman with smooth, shiny, thick copper-colored hair, clear, intelligent eyes, and an innate elegance and grace. She was strong, courageous, proud, and had a sense of honor and family that I have rarely encountered in others. She "adopted" Luca first, then me, when we arrived in Kenya: she was like our African mother, always ready to give us advice and help us when we needed it.

When I was going through a crisis or needed a break from the heaviness of Africa, I would run to her. Her home was beautiful and welcoming, always open to friends, just like her heart. It was surrounded by a large garden, where beautiful trees stood and colorful flowers grew, which Picci tended to with infinite love and dedication. The grass in Naivasha was always green, thick, and lush because it rained often. During the day, the sun scorched, making it impossible to be outside without a hat or shelter; but at night, it was cold, very cold indeed, because the lake was situated at an altitude of 2,000 meters.

In Naivasha, every morning, we would walk with Picci on her farm, among rows of green, tender beans planted with the utmost order and precision. Kikuyu women, with their brightly colored clothes and headscarves, bent over with their legs perfectly straight, a posture only African women seem capable of, picked the vegetables, cheerfully singing their melodious songs. Picci had also planted orange trees in memory of her beloved Sicily. When we entered the orchard, picking ripe oranges and lemons and breathing in the scents of distant summers, it felt like being in a place outside of time, in the old Sicily described in "The Leopard," a novel I have always loved dearly.

Over the years, we often visited Picci. After a few days in her company at Naivasha, we always felt safe and protected, happy and at peace with the world. We would return to the Chyulu Hills with a light heart and a deep serenity that lingered for many days. We were also closely connected to her daughters: Francesca, Valentina, and Simona. Beautiful, kind, and strong-willed women who deeply loved their mother and Africa, they became dear friends from the very beginning of our African adventure. I particularly bonded with Valentina, whose character I found very similar to mine—always cheerful and ready to help. Such was the close bond we had with our Naivasha family, that Picci stood as Luca's witness at our wedding, and Valentina as mine. Ours was a communion of souls, often transcending the bounds of ordinary friendship.

In Nairobi lived Renzo and Annalisa, dear friends who had been in Kenya for over 30 years. Renzo was born in Africa and met Annalisa in Italy, during his military service. They married in Italy and moved to Nairobi shortly afterwards. They had three children. Whenever I needed to go to Nairobi before the birth of my children, they opened their beautiful home in Karen to me and hosted me until the day of the delivery. Annalisa and I kept in touch frequently and became close friends. She was always kind, practical, and willing to help. Whenever I went to Nairobi on errands, we would meet, and she would fill me in on the latest news from the social scene in "Happy Valley."

These were just some of our African friends. It did not matter that they came from old aristocratic families, pioneers, or simple farmers who had arrived from Europe or elsewhere many years ago, while we had come from Italy only recently. Our backgrounds could not have been more different; yet we were united by the same boundless love for Kenya and its people, and the certainty that we could never live anywhere else. We shared a deep gratitude for this extraordinary country and a poignant longing to spend our final days here. Perhaps this is why such strong bonds form between those who live in Kenya, even if they are hundreds of kilometers apart. Perhaps this is why friendship takes on such a profound meaning for those who, like us and them, choose Africa as their new home.

Arusi (wedding) in Timau

I had just opened the beautiful white envelope, made of fine, elegant paper, that had arrived in the mail; I pulled out the invitation and exclaimed: "How wonderful! Oliver is getting married!"

Oliver was the firstborn son of Graham and Vera. He had met a lovely girl while at college in England. After being engaged for two years, the young couple had decided to marry. The ceremony was to take place in Timau, and the bride would move to Kenya, to live on the MacGregors' estate.

A wedding in Kenya is always a grand affair, meticulously planned and brilliantly organized by the bride and groom. It is a chance to reunite with dear friends who live hundreds of kilometers away and are seldom seen. As in Europe, it is a moment of joy and love; but in Africa, it always carries that "something extra"– something magical that makes it truly unforgettable.

I noticed this on my own wedding day and at every other wedding I attended in Kenya. They all, without exception, had something wondrous and enchanting about them. It was the atmosphere, more than anything else, that made everything so beauti ful. Kenyan weddings were always held outdoors, in unique places, surrounded by friends who sincerely shared the couple's happiness, not just out of formality! I knew it would be the same for Oliver and Elizabeth's wedding, and I could not wait to attend.

Finally, the big day arrived; it was a beautiful morning in February, warm and pleasant.

We had set off early, flying north in our airplane. When we landed in Timau, on Graham's private grass strip, there were already about a dozen small planes parked along the sides. This was another thing I loved about weddings in Kenya. Almost all the guests arrived in their own planes. You would see them approaching like colorful birds, wings gleaming in the African sun, flying low over the ceremony site to announce their arrival to the soon-to-be-wedded couple.

Flying in Kenya is like driving a car. Almost everyone has their own plane and uses it regularly. Having one is not a sign of luxury or excessive wealth, but simply practicality and normalcy. The roads are dangerously full of trucks and are often impassable. It could take hours to cover a distance that, by plane, could be flown in just a few minutes.

I loved seeing so many planes gathered on a single grass strip in the middle of the bush. Knowing that they belonged to people who lived far away and had flown for hours to celebrate with their friends filled me with joy and made me appreciate Africa even more.

As soon as we landed, a Land Rover came to pick us up and took us to our campsite. There were over two hundred guests at the wedding, and most, like us, would spend the night in tents and return home the next day.

A “fly camp” was always exciting, and this one had been organized to perfection.

It meant spending the night in a tent, surrounded by the sounds of African nature, accompanied by the magic of the moon under a roof of bright stars. It was the perfect balance between romantic and adventurous.

Each guest tent had two beds with freshly laundered sheets and soft blankets, two towels, a nightstand, and a flashlight. On the veranda outside, there were two safari chairs, a small table, and even a mirror so the ladies could do their makeup before the ceremony. Outside each tent, a basin of clean, warm water was brought for washing or shaving, but the real “luxury” was that there were hot showers and pit toilets with all the comforts for those who needed them!

Our campsite, located on a vast plain with an incredible view of Mount Kenya, had over 50 tents set up. Yet they had been perfectly arranged, allowing each one its own privacy. As we arrived, we greeted many friends we had not seen in a long time. We settled into our tent, took a shower, and dressed for the ceremony.

It was amusing to see the ladies entering their tents in safari clothes or bathrobes and emerging half an hour later in elegant, long dresses adorned with sequins and chiffon scarves, hair and make-up perfectly done! They often walked barefoot or wearing flip flops, with a pair of high heels underarm, which they would only wear upon reaching the ceremony site, smiling on the arms of their gentlemen, who were in tuxedos, looking elegant and utterly at ease, as if it were the most normal thing in the world.

Oliver and Elizabeth had decided to get married in the late afternoon, in an enchanted place: a forest of African cedars. The path to the ceremony site was lined with oil lamps and strewn with rose petals beneath our feet. We walked it in silence, admiring the beauty of the landscape around us and trying to fully absorb the mysticism of the moment.

Every member of the couple's family and their closest friends had helped with the planning. Some had marked trails and built wooden steps in the forest to make the ceremony site more accessible to everyone. Others had created natural arches from wooden trunks decorated with moss, lichens, and wild orchids, and even constructed a small wooden bridge over a stream, its edges adorned with fragrant wild jasmine.

After about 10 minutes of walking, we reached the spot chosen by the couple for the wedding. It was a marvelous place, magical and deeply spiritual. Among towering cedar and acacia trees, their branches entwined with garlands of moss and fresh flowers, a clearing had been created. In its center stood an altar, a massive single piece of wood the height of a table. It must have come from an ancient tree. You could tell by the roughness of its bark and its incredible diameter. On it, there was a wooden cross and a few fresh flowers. A natural cathedral had been created in the middle of that forest: it was holier than any church I had ever visited. Bales of hay had been arranged in front of the altar for the numerous guests to sit on.

But the most astonishing thing was a magnificent grand piano under a tree, near the altar. It was being played by an elegant elderly English lady in a sky-blue chiffon dress, her wide-brimmed hat adding to her graceful presence. She began the ceremony by impeccably playing the British national anthem. A children's choir from Nanyuki had been invited, and with their sweet voices, they sang sacred hymns throughout the ceremony, touching us all deeply.

The MacGregors have Scottish roots. Thus, the groom wore a traditional kilt, which suited him perfectly and made him proud. I thought about how quickly he had grown, what a fine man he had become. When I had first met him 10 years earlier, he had been a shy and reserved boy. And now here he was, ready to start his life with a beautiful girl, ready to face the future with courage.

When the bride arrived, in her long silk organza gown with warm orange hues, holding a bouquet of white and orange calla lilies in her hands, everyone turned to admire her. Elizabeth walked slowly, smiling, on her father's arm, accompanied by six bridesmaids holding small bouquets of fresh flowers and wearing white flower crowns on their heads. I watched as Oliver looked at his bride with pride, and I glimpsed a shadow of emotion in his blue eyes. She returned his gaze with a smile full of love and sweetness. At that moment, I understood how much they loved each other, and I squeezed Luca's hand tightly in mine.

After the ceremony, we made our way to where we would dine and dance. Waiters in white gloves began circulating among the guests, offering glasses of champagne and hors d'oeuvres. In the middle of a plain, among impalas, zebras, and gazelles curiously watching us, ochre-colored Moroccan tents had been set up. In front of them, two enormous bonfires were lit to warm the cold night at the foot of Mount Kenya.

Under the tents, tables were laid out to seat all the guests for dinner. They were impeccably set with long white linen tablecloths and decorated with scented candles and fresh flowers. Brass Moroccan lanterns added a unique touch to the scene. In one corner, antique kilim rugs had been laid on the ground, along with large Indian cushions, so guests could sit down and relax after dinner. Crackling braziers added to the festivity and warmed the air. The tents were arranged in a horseshoe shape, and in the center, a wooden platform had been set up, where we would dance until dawn. All enjoyed the dinner: Moroccan cuisine, but with English desserts, of course!

The dancing began, led by the radiant and emotional bride and groom. The music was fantastic, and we danced for hours under the African stars. Late at night, exhausted and shattered, Luca and I went to sleep in our tent, holding hands.

The next morning, a brunch was set up for all the guests. After a hearty breakfast, chatting with friends and reminiscing about the night before, we set off again for the Chyulus. With the roar of the guests' planes flying low over our heads, we went to bid farewell and thank the newlyweds.

Another wonderful wedding had been celebrated, and we had been privileged to witness it.

An Unexpected Gift

Giovanni and Elena were extraordinary friends, remarkable travelers of authentic, unexplored Africa. I will always be deeply grateful to them. Through their pure and profound love for Kenya, we received an unexpected gift: a splendid house in Malindi, a coastal resort that rose to fame in the 1970s, nestled on Kenya's coast, kissed by the warm waters of the Indian Ocean.

How can something this crazy happen? It makes no sense...

Well, it is all related to the love for Africa and to what Africa can do to you...

From the early 1980s, Elena spent at least six months each year in Malindi, and Giovanni visited whenever work and commitments in Italy allowed. Giovanni was a prominent and successful lawyer as well as being a talented writer, a lover and collector of unusual objects, such as antique toy soldiers, fine prints, rare books, and a connoisseur of wines and gourmet foods. A captivating man with a brilliant mind and a big heart. Elena was a beautiful and elegant woman, possessing great taste and refinement. Having married Giovanni at a young age, she had always followed him on his adventurous journeys, indulging his eccentricity and every whim.

Over the years, they had come to Campi ya Kanzi many times. Having them at the camp was always a pleasure; their tales of travels and deep reflections on stimulating topics kept us company on magical starry nights. Yet lately, we felt their worries. Aging without kids of their own made them think what would become of the magic place they had created in Malindi, a house they loved, and which made them live their African dream.

One day, while walking on the beach of our beloved Kiwayu, we heard a familiar voice calling us. Unbeknownst to us, Giovanni and Elena were also there.

They saw us walking on the beach with our kids, Lorenzo was only a few months old.

I guess in us they saw their younger selves and they understood that we were seeding our love for Africa into a new generation. A generation of kids born and raised in the bush, cradled by lions' roars as their lullabies.

That evening, sitting on the white sand, under a dramatic moonless starry sky they said to us: "Luca, Antonella: your love for Africa has meant so much to us. What you have created, your love for the Maasai, your protecting their culture, their land, their wildlife heritage, means the world to us. We see in you and your dedication to Africa the same feelings that inspired us decades ago. We are at our dusk, and we wish our love for Africa to continue into a new generation. We would like our home in Malindi to become your home, so you can use it to benefit your foundation and keep impacting so many Maasai lives".

We were shocked, speechless and in disbelief.

The next morning, they left, and we did not think much about the odd conversation we had had the night before.

Two months later Giovanni called and asked to meet up to make the donation legal.

They had decided to donate their house to the foundation that we had created: "The Maasai Wilderness Conservation Trust."

Our friends gave us a truly "unexpected gift" that we accepted with disbelief and emotion. As the new owners, we naturally assumed all the costs of ordinary and extraordinary maintenance. We also committed to finding paying guests who, before or after their safari at Campi ya Kanzi, would want to extend their vacation for a few days, immersed in the peace and magical beauty of a splendid private villa, just a few meters from the ocean in Malindi. In this way, the house would generate new income for the Foundation, and Giovanni and Elena, according to their wish, would continue to help the Maasai and thus the Africa they had so deeply loved.

When Giovanni passed away just a few years later, we told the story to our kids and to Samson, the foundation chairman, and to his kids.

Adults and youngsters were all reduced to tears. How inspiring and what a legacy...

We first visited the house in December 2011. After just an hour's flight from Campi ya Kanzi, we landed in Malindi. We took a taxi from the airport, and within minutes we arrived at the house: "Nyumba ya Kipepeo," was written on the wooden gate. "The House of Butterflies", in Kiswahili.

The gate opened onto a galana stone avenue, lined with tall coconut palms, aloes, and agaves. Kanyua, the young woman who had always taken care of the house for our dear friends, came to meet us; smiling and cheerful, as most African women usually are. We entered the large house almost on tiptoe, not wanting to disturb.

We could hardly believe that this house was ours. The house was beautiful, silent, spacious, with large windows overlooking a stunning tropical garden. Soft white gauze curtains, billowing with the warm ocean breeze, adorned the veranda's arches, lending elegance and charm. The atmosphere was relaxing and tranquil. Elena had decorated it with great taste and refinement, in every detail, from Persian rugs and Afghan kilims to fine furniture. Between exquisite African statuettes and Giovanni's rare antique books, one could feel our friends' admiration for beautiful things and their dear Africa. Every object in that house told a story: a story of passion and dedication, a story of hope and life.

The "Butterfly House," as I affectionately named it, swiftly became a dear companion. I stood in its presence, quietly, sensing its slumber, as if it were a living being, breathing slowly, yearning to awaken, to feel alive once more. For three long years, it had been untouched by the warmth and love of our friends, who once made it vibrant and full of life. Now, it lingered in a sorrowful stillness, longing to be cherished, to be brought back to life. And here we were, the ones destined to breathe new life into its silent halls, to love it as it deserved. I could already see it in my mind's eye—our family, our children, our friends, and the guests of Campi ya Kanzi—all gathered in this sanctuary.

I imagined my children running barefoot across the lush grass, their laughter filling the air as they swam and dived into the pool, cycling along the galana paths shaded by towering palms, and gathering vibrant flowers from the garden's embrace. Their joyful voices would echo through the wide rooms of the Butterfly House, a melody of life and love that would resonate in every corner. I was certain that Luca and I, with our boundless love and passion for Africa and its people, would infuse this beautiful house and its enchanted garden with energy again. The house would soon awaken from its slumber, vibrant and alive once more, cherished and essential, a place of warmth and welcome.

The Butterfly House also boasted a beautiful garden, lush and quiet, surrounding and protecting it from the perils of the frantic outside world. It was enormous, alive, and precious, made to bring peace to the heart and breath to the soul. This garden, with its magnificent colors, intoxicating scents, and a whispering wind through the branches of ancient baobabs, was an enchanted place.

I saw myself walking in that beautiful garden every day, among dozens of tall and elegant palms, the path adorned by ancient baobabs, blooming oleanders, acacias, desert roses, mangoes, papayas, bananas, frangipanes, and lemon trees. And joyful splashes of color, scattered here and there on the perfectly green grass, like paint on an artist's palette: bushes with large fragrant flowers in yellow, red, white, pink, like delicate goblets of fine Venetian glass, opening every morning, renewing the wonder of nature. And the beautiful bougainvillea in pink, orange, white, and purple that covered the coral wall, blessing it with vibrant hues.

At the bottom of the garden sat a small but lovely pond where placid and gentle white-water lilies floated, opening shyly at dawn and closing modestly at dusk, as if to protect themselves from some unknown danger. And the tall, strong acacia by the pond, home to hundreds of nests of industrious and tireless yellow weaver birds, like a Christmas tree decorated with dozens of baubles made of twigs and dried leaves: their nests. And the delicate colorful butterflies, flying freely, gracefully alighting from flower to flower. And the tadpoles and frogs, croaking peacefully in a melodious nocturnal concert. And the darting little fish, swimming swiftly in the clear waters, and the green and orange dragonflies, tirelessly buzzing low over the water. And the slender, unruly papyrus bushes, surrounding the pond with care and diligence.

At dawn, the sound of the ocean would wake me from my slumber, the wind riding its swelling waves, echoing all the way to the house. Then began the chorus of the songs of dozens of little birds, gentle inhabitants of the enchanted garden, joining in a single melody. In a song of Nature.

And then, again, to lie down in the fine, fragrant green grass, close my eyes, and be moved by a subtle nostalgia for our past life, me as a child at the Castellino or in the countryside house where Luca and his family lived in Italy, in Castenedolo. Before everything, before Africa. To relive the deep emotions of those distant times, and to feel the love and presence of those who had passed once more. And immersed in the peace of Africa, to realize that we are our homes, and wherever we go, we can relive what is deep in our hearts, whenever we wish, no matter where.

After months of inner fatigue and a search for my spirit and energy, I found them in the most unexpected place of all. In Malindi, in Giovanni and Elena's house. Kiwayu was, and remained, in my heart the wild and fascinating place it had always been. It would still be our adventure, our dream that seemed like reality. But Malindi, with its beautiful energy and its Butterfly House, gentle and generous, welcomed us with open arms, like a mother who, after years of waiting, sees her lost children return. Malindi was something else: it was like returning home, to rest, safe, calm, and at peace. The place from which you would never want to leave.

When I walk in the savannah and see a hawk soaring in the sky,

I think of Giovanni. I am honored and grateful he saw in us his love for Africa.

I want to believe he is happy to see how his beautiful house has brought joy to us and support to the Foundation and to the Maasai.

Thank you, dear friends, for the unexpected gift of your love: thank you, Africa, for what only you can give to the hearts of those who understand.

African Childhoods

Lucrezia, Jacopo, and Lorenzo are children of Africa. Their bond with Kenya and its people runs deep, a connection both physical and spiritual, as if the very land pulses through their veins.

The first words my children spoke were in Kiswahili. Only later did they begin to converse with me in Italian, and eventually in English as they entered school. Today, their primary language is English, though they understand Italian perfectly. Out of habit, they speak it less, but I insist on addressing them in Italian, even if they reply in English! When my children return to Kenya for their school holidays, they seamlessly resume speaking Kiswahili with the Maasai, as if it were their mother tongue. This fills my heart with joy.

From an early age, we took our children to Italy at least twice a year. We wanted them to connect with the world from which their parents came, to love and cherish their Italian heritage as well. In this way, I believe, they have absorbed the best of both worlds. Comparing them to their Italian peers, I see them standing apart – wiser, and more mature. I wonder if this is because they have been exposed to experiences so different. Their early years, spent entirely in the purity of African nature, in the company of the Maasai and surrounded by the magnificent wildlife, have undoubtedly left an indelible mark on their souls.

In them, the brightness, spontaneity, sense of freedom, and indomitable spirit of Africa coexist in perfect harmony with the strength, independence, and pride of the Maasai who raised them. They also carry within them the sweetness and passion of Italy, an innate respect and appreciation for its ancient art and culture, a love for its superb cuisine, and that quintessential Italian knack for getting on with things! My children embody all of this, most probably thanks to the genes of my husband and I, of their grandparents, and all those who came before them. Yet I like to believe that Africa, the Maasai, and Italy have played a fundamental role in shaping their character and strong personalities.

When Lucrezia was just five days old, we took her to the camp from Nairobi, as we did with Jacopo and Lorenzo. But Lucrezia, unlike her brothers, happily lived with us in a tent until she was two years old. I believe that living in a tent for so long profoundly influenced her, in ways we, as parents, could scarcely comprehend. Her love and passion for all animals and nature seem inextricably linked to her early and intense connection with them. I remember one night, lions were roaring outside our tent, waking Lucrezia. She must have been around eight or nine months old. She responded with a gurgle, a sound resembling theirs, and then, smiling, she drifted back to sleep as if hearing lions roar just a few meters away was the most natural thing in the world. I found it extraordinary then and still do now. I was certainly not as calm as she was that night! The same was true for her brothers: they always showed complete calm and fearlessness when the lions' roars filled the African nights in their earliest days.

The first time Lucrezia saw a lion with its full mane, she was two years old. We had taken her to Okoikuma for a walk, and on the way back to the camp, we spotted the proudest of animals – an immense, regal lion walking, calm and unperturbed, right in front of our Land Rover. Lucrezia, thrilled, exclaimed, "*Simba, simba!* Lion, lion!" It was the first time she had seen one up close, and she was not the least bit frightened! The lion did not even glance at us, continuing its stroll for about 10 minutes, unbothered and deep in thought. It was sunset, and he was probably beginning his hunt. We followed him in silence, at a walking pace, careful not to disturb him. Finally, he decided to let us pass, moving aside. Lucrezia was disappointed to lose her new friend so soon and asked about him for several days.

Even today, though we live in a comfortable house, my children's favorite activity is to sleep in a small tent, in the bush behind our home. When they were younger, Luca would go with them. In the evenings, they would prepare carefully for their adventure, packing everything they needed for the night in their backpacks, never forgetting to bring a book to read and a flashlight. Then, hand in hand with their father, they would head out excitedly into the starlit African night to face their "great adventure." The following morning, they would return home, their eyes sparkling with joy from an exhilarating night, eager to share tales of the animals they had heard near the tent and the birds that woke them in the morning.

I remember one night when three lions circled the small tent where Luca, Lorenzo, and Jacopo were sleeping. I stayed awake until dawn, ensuring nothing happened, worried and scared. I had called Munghesia and Daudi, the askaris, who stood guard outside the little tent all night while the lions roamed around... Thankfully, nothing happened: Luca and the kids slept soundly through the night, completely unaware. But it was a sleepless and exhausting night for me!

Sometimes, on nights where the full moon shines brightly like a pearl in the sky, we would see a lion come to drink from the small waterhole under Lucrezia's bedroom window, a pond we created for the animals and birds. My children often tell me they have been woken at night by the sound of a lion drinking, its tongue lapping the water and for them, it is a normal occurrence, part of their daily life.

In the afternoons, at sunset, we sometimes take a walk or drive behind the house to have tea with our children. It is the most beautiful and precious time of the day, with the golden light painting the Chyulu Hills, the sky glowing red and orange, and the sun sinking behind Kilimanjaro. We settle on a Maasai *shuka* under our favorite acacia, unpack steaming thermoses of tea and milk, and ginger biscuits made by Patrick, savoring every moment we spend together, enveloped in the magic of these precious times.

We watch the curious giraffes as they approach, and the gazelles that flee at the slightest movement or sound. Our dogs chase each other joyfully, and Jacopo and Lorenzo delight in teasing me with strange insects hidden in my tea! We always return from these outings with joy in our hearts and peace in our souls, feeling at one with ourselves and connected to the extraordinary nature of Africa.

When Lucrezia turned three, we found an American girl, a kindergarten teacher, to come live at the camp to teach her along with some Maasai children. Living an hour's flight from Nairobi, it was impossible for us to take her to school every day. Clare was 23 and came from Seattle. She was a beautiful girl with curly blonde hair, bright blue eyes, and a radiant smile that lit up her lovely face. She quickly adapted to life in the bush. Her simplicity, sweetness, and ability to adjust made her a perfect fit for our extended family of Italians and Maasai. She was supposed to stay only a year, but her attachment to all of us inspired her to extend her stay, and Lucrezia quickly grew to love her dearly!

We built a small school at the camp, the Kanzi Academy, to provide our kids and some Maasai children with an excellent education. We equipped it with desks and chairs, blackboards, stationery, computers, and even a DVD player to show cartoons or interesting documentaries to the students. Outside our little school, we put down sand with wheelbarrows, buckets, and spades, and built slides, swings, and other playground equipment to entertain them. But their favorite thing was the trampoline we gave Lucrezia for her fourth birthday. It was big enough for all the children to jump on together. Lucrezia, her little brothers, and the Maasai loved it dearly and spent countless happy hours bouncing and laughing heartily!

All three of my children attended the Kanzi Academy with their Maasai friends until Lorenzo turned 10. Our little school in the savannah, of which I was the proud director, remained open for 14 years, during which many teachers from all over the world came and taught with passion and dedication. I created a school curriculum that combined regular academic activities with transcendental meditation, yoga, art, drama, and teaching methods inspired by Montessori and Steiner. This holistic approach allowed the students to develop their artistic, creative, and intellectual abilities while respecting the magnificent African nature surrounding them.

As the children grew older and needed more interaction and socialization with others their age, we decided to send them to school in Nairobi. I enrolled my children in an international school in Nairobi for a couple of years before they went to a boarding school in Switzerland to finish high school. Lucrezia struggled with homesickness and missed her beloved Africa when we sent her to Europe to study. For Jacopo and Lorenzo, it was easier; perhaps because they were together, they quickly made friends and adapted well to their new world and their new school near Geneva.

I believe that the Kanzi Academy has shaped my children into compassionate and respectful individuals, deeply attuned to the nature, people, and animals surrounding them. There were times when Clare, and the other teachers who came over the years, had to interrupt their lessons because word would reach us of a lion or cheetah near the camp. On such occasions, we would hop into the Land Rover and take all the children on a safari to see the animals! I am certain they learned much more from these experiences than they would have in a morning of classroom lessons!

Today, what brings us the most joy is witnessing the strong bond of kinship that has formed between my children and the Maasai children (now boys and girls) who grew up with them. When they were little, they played, sang, danced, and ate together, like brothers and sisters, noticing no difference in the color of their skin or the worlds from which they came. They were, and still are, all children of Africa – *Watoto wa Africa* –and that is what truly matters.

Raising children in Kenya is, I now realize a great privilege, especially at Campi ya Kanzi, surrounded by untouched nature and immersed in complete freedom. The climate, the clean air, and the lack of pollution make our paradise a truly wonderful place to live and to raise our children. What was absolutely normal for Lucrezia, Jacopo, and Lorenzo in their early years – like following a lion's tracks on their way to school in the morning, after hearing its roar outside the house the night before, or being surprised by a giraffe peering curiously through their bedroom window, or taking a Sunday morning horseback ride among gazelles, warthogs, and zebras—was, and still is, unimaginable for most of the children of our friends living anywhere else!

Kenya – and especially Nairobi, its capital – is a very cosmopolitan place where different races and cultures – Indians, Asians, Arabs, Europeans, and Africans – coexist in complete harmony. For children growing up in Kenya, it is normal to interact with different cultures, and they quickly learn to accept and appreciate diversity, which greatly broadens their mindset.

Sometimes I wonder if I made the right decision in raising my children in Africa. One certainly cannot say that my children have had a "normal" life or one similar to what I was accustomed to as a child. But then I would look into their eyes, so full of serenity, joy, curiosity, and love for how they were living and what surrounded them, and my motherly worries would be set to rest.

Today, I look with pride at the way my daughter Lucrezia cares for, helps, and respects those around her, whether they are like her or different, with a natural ease that sometimes makes me envious. Her desire is to return to Kenya after completing her university studies and work with the Maasai on the many conservation projects managed by our foundation. I am certain she will do an excellent job.

I look at Jacopo, with feeling and admiration for his generosity and love towards all living creatures. Jacopo's dream is to become a doctor, like his grandfather, and open a hospital in the savannah where we live to treat and help the Maasai community there. I am sure he will be a wonderful doctor, and that his grandfather Guido would be proud of him.

I look at Lorenzo and am so impressed by the strength of character and determination he shows every day. At 16 years old, Lorenzo is already capable of flying our plane. His father began teaching him when he was a little boy, and today, my son can take off and land our plane on his own. Flying has always been his passion, and becoming a skilled pilot is what he wants to do when he grows up, so he can soar over the vast African plains. And I am certain that only the sky will quite literally be his limit.

My children are the meaning of my life: they are children of Africa, and I am so happy that they are!

Mother Africa

In time, and with great effort, I have learned to know and love Africa.

At first, I rebelled against its harshness, its ever-looming presence, so wild and untamed. I could not accept the idea that I couldn't dominate it, that I had no power over it, as I had been taught in "my world." But gradually, gently, I surrendered, and with that came an understanding, and a deep love for this ancient, wise land, so unique and so special.

Africa asks everything of us, but in return, it gives us everything.

In Africa, we are laid bare, and we are ashamed of our nakedness. There is no hiding; we must rely only on our bodies and souls, on our real selves. Everything else is irrelevant. The certainties we held in the First World become doubts, hesitations and fears in Africa.

Slowly, we realize that our senses are not enough to help us survive. In Africa, the light is blinding to our eyes, the silence deafening to our ears; the mighty roar of the lion is too intense to bear without awakening primal fears within us. The red earth is too rough when we hold it in our hands, the heat is unbearable, and the sky is so vast, so infinite that it makes us feel insignificant and small before the power of Nature.

In Africa, we strip away all that is form and pretense; everything we are not nor do not have sheds truth on all we believe to have possessed. We find ourselves alone with our fears and our limits, and we finally understand that we can only rely on ourselves to overcome them and continue living—or perhaps, to truly start living.

And then, if and when we overcome the fears and the physical and mental limits that we believed were inseparably part of us, when we become aware that we have gone "beyond," that we have grown, that we have taken that extra step; then we feel proud, free, and so at peace with ourselves that nothing else matters or can shame us any longer. Then, nothing can frighten us anymore.

This is the reward that Africa had in store for me. This is what the wild Africa grants us, and nothing else can be more valuable than this prize, for it returns us to ourselves.

It is the price we must pay to know Africa, a land that can give us such overwhelming emotions that our hearts suddenly burst with feeling. A land that can bring us to tears before its beauty and vastness and the sweetness and infinite goodness of its people. But also, a land that can take everything away from us in an instant, without reason or explanation, and without a second chance.

Perhaps it is for this reason that one becomes so emotional and intense in Africa. Perhaps it is for this reason that in Africa we feel our hearts beating every day, and we feel alive as in no other place in the world. Perhaps it is for this reason that, to me, now, Africa is home, and I decided to marry here and to give birth to my children here. Nowhere else in the world would I feel so at peace with myself, so sure of doing the right thing.

Africa, at times, can seem harsh and cruel. And it is, because it teaches us to be ruthless, even with ourselves, in order to survive. In Africa, we cannot afford to live in the memories of the past or the plans of the future; we live in the present, in the moment, so intensely that it fills our entire minds and hearts, leaving no room for anything else.

I believe that this is how it should be for everyone. I am learning from Africans to live with what I have and to appreciate the fact that I am alive today.

In Africa, we must learn to accept that a child we have always seen playing around us might suddenly fall ill and die, and there is nothing we can do to help. Unfortunately, doctors and efficient hospitals are not present in the remote area where we live, and sick people must go to Nairobi, 350 kilometers away from us, to be properly treated. So, despair overtakes us, and the most terrible doubts creep into our minds: "Perhaps, if I had realized earlier, if I had understood, perhaps then I could have done something for that child, I could have saved him, I could have..." and our hearts scream against such a tremendous injustice.

But then we look at the land that surrounds us, so beautiful, so mysterious, and the sweet serenity in the eyes of its people; and in the mystery, we accept both Death and Life. Africa teaches us humility; it teaches us to accept that we are not the masters of existence.

One day, when we were building the camp more than 25 years ago, a Maasai man came to me after walking all night and told me that his wife was very ill, perhaps dying. I said I would take her to the doctor in Nairobi, because I didn't know what to do, nor was I capable of treating her. He replied that there was no need to go to any doctor, and that surely, I could save her. When I asked why, he said: "Because you are white."

Sometimes I wonder: "How can we 'white people' meet these expectations and successfully bear the weight of such a great responsibility? What skills do we truly have to earn this reputation?" In the early years in Kenya, while we were building Campi ya Kanzi, every morning a long line of Maasai men and women would wait for me to treat all their ailments. They had nicknamed me "Mama Daktari." My knowledge of medicine was rudimentary to say the least, but I tried to do my best, where I could, to alleviate the symptoms of their illnesses. I was the only person for miles around who had any medicine, and so, despite myself, I quickly became a precious resource for the Maasai who worked with us. But I felt the weight of a great responsibility in doing so and I wondered every day if I truly deserved the trust of the Maasai.

Fortunately, today, through donations to our foundation, we have built a dispensary and employ a doctor and several nurses who take care of the Maasai living near Campi ya Kanzi. This has helped a lot in making the life conditions of the local community much better and acceptable.

Although they often ask us for advice on how to solve situations that are unfamiliar or complicated for them, unlike many other African tribes that have westernized over the years, the Maasai never forget who they are. They are so proud of their traditions, their customs, and their culture that sometimes they may look at us with superiority and detachment.

All attempts by those who have come to Africa from the western world to make them like us have been in vain: in their hearts, they always remain true Maasai. This is what perhaps makes them so special, so untamable, so unreachable, fascinating, and unique.

This is also the reason why Luca and I decided to create the Maasai Wilderness Conservation Trust (MWCT).

We oversee a variety of projects, from employing rangers to combat poaching and the killing of lions, an endangered species, to building schools and dispensaries. We protect and preserve the precious cloud forest on the top of the Chyulu Hills. We employ preschool, primary, and middle-school teachers, as well as a doctor and nurses, to provide minimal healthcare. It is simply incredible to think back to the beginning of our African adventure more than 25 years ago and to see what we, and those who believed in us and generously decided to support us on our long journey, have accomplished. Today, the foundation employs more than 400 people, all locals. It has received numerous awards, both nationally and internationally.

Our hope is to improve the quality of life for those who live alongside us and to give the Maasai children the opportunity to receive an education so that one day, they can choose how to manage their land and how to live their lives. We wish to give them the tools to become the leaders of their future and thus be able to preserve their land, and the animals that inhabit it, in a sustainable way.

Our mission is to ensure that these beautiful people of Africa and the land where they live are safeguarded and maintained in their authenticity and naturalness.

There is much to be done, but much has already been accomplished, thanks to the generous help we have received. We are hopeful and grateful, and we will continue to fight to preserve this untouched corner of Africa and its proud inhabitants.

Naseryan

My feelings for Africa are interwoven with stark, intense contrasts, just like Africa itself. Africa is blinding light and complete darkness, life and death, the most intense joy and the deepest sorrow. There is no middle ground or half nuances. Nothing is "gray," only vibrant colors that are alive and intense.

There have been moments during my life here when I felt so happy and grateful to be in Africa that I was almost afraid of being too fortunate to deserve all this. Other times, Africa has taken everything from me, suddenly, leaving me desolate and empty. But I have never regretted, not for a moment, my choice to live here.

When I first arrived in Africa, I was just Antonella to the Maasai. Now they call me *Naseryan*, "the one who helps."

Never in my life have I felt such a weight of responsibility toward others, never have I felt so obligated to be consistent and serious. Since I was a girl, I have always detested being what others expected of me; I could never stand having to live up to others' expectations.

But this time, the "others" are the Maasai, who love and respect me like a mother, who help and protect me as if I were their daughter, who accept me every day as if I were their sister.

Now, it is different. It is in being "the One Who Helps" that I find fulfillment every day, that I selfishly draw my satisfaction, that I am, at last, myself. It is in the deep, laughing eyes of Maasai children that I draw my strength when I feel tired and drained. It is in their proud smiles that I renew myself and live another day in Africa.

I do not know what will become of us and Africa in the future. "Future" is too big a word and holds little meaning in the world I live in now. Only one word is part of my being now: Love.

Love for Luca and my children.

Love for Africa, its people, and its animals.

Love for what we all come from.

If we want to preserve this ancient continent and its inhabitants, we must act swiftly. In 50 years, what surrounds us might no longer be here. The lions, the majestic kings of the savannah, the elephants and the rhinos – they are all disappearing, giving way to crops that feed an ever-growing and now overwhelming number of human beings. The same is happening to many other species of African animals and native plants. This depressing and not-too-distant future seems unstoppable and inevitable.

If we let it all vanish, we will bear a great responsibility to those who come after us. What Africa will my children see?

And their children, your children – what world will they find?

It is with this unsettling question that I put down my pen.

I have only one certainty now: Love conquers all, and Hope is the child of Love. But sometimes, Love and Hope are not enough...

Twenty-Five Years of Africa

EPILOGUE AND ACKNOWLEDGEMENTS

"Sei morta, finita, uscita per sempre dalla mia casa e dalla mia vita! Non sentiro' più il tuo muso umido cercare la mia mano per una carezza, nè sentiro' più la tua zampa appoggiarsi alla mia gamba per cercare anche solo uno sguardo nei tuoi occhi ormai quasi ciechi. Non seguiremo più, ansanti, le starne sulle colline nè cercheremo più la speranza di una beccaccia d'autunno. Altri cani si accucceranno ai miei piedi, davanti al fuoco, ma non sarai mai più tu. Mio figlio, che ha imparato a camminare aggrappato a te, è sempre più grande e robusto e mi segue a caccia, ma presto, forse troppo presto, se ne andrà solo, con un'altra Cochi, nei boschi sulle colline; quando i miei scarponi non cammineranno più."

"You are dead, gone, walking out of my house and my life with no return!

I will no longer feel your damp nose asking for a caress from my hand, nor will I ever feel your soft paw on my leg searching for even just a glance in your eyes, now almost blind.

We will no longer follow, panting, partridges on the hills, nor will we hold onto the hope of an autumn woodcock.

Other dogs will nestle at my feet, curled up in front of the fire, but it will no longer be you.

My son, who learned to walk gripping onto you, is taller and stronger every day, and follows me on my hunts, but soon, perhaps too soon, he will walk alone, with another Cochi, in the woods on the hills; when my boots will no longer walk."

This is a poem by my father-in-law, Guido Belpietro, written for his dog Cochi, whom he adored and who left him too soon. Cochi was Luca's first dog, given to him when he was just a child. They lived in symbiosis; she would loyally protect him and love him unconditionally, like only a dog's heart can. The first night in which Luca, at only four years old, slept in a tiny tent at the bottom of his garden in Italy, Cochi was right by his side to protect him and watch over him until dawn. If Luca had found the courage to sleep alone in a tent at such a young age, it was to show his father that he was ready to follow him in his African safaris. And it was also because, with Cochi by his side, he knew nothing bad could happen to him. Luca's infinite love and passion for Africa was born on an Italian summer evening, beneath a starry sky, in a tiny tent at the bottom of a vineyard, where a young boy held his dog Cochi after having listened attentively to the tales of his father's adventurous safaris.

My only regret is not having known Guido better. I remember him as a man full of love and life. He was constantly surrounded by good friends and electric energy. I spent only a few months with him, before his premature death. He loved Mina's songs just as Luca and I did. I would play my guitar for him, and we would sing together. Those were moments of profound emotion shared between us, which I look back on with a bittersweet nostalgia. My father-in-law passed away in August 1993 before Luca and I decided to begin our African adventure. He had always dreamt of living in Africa and would have been thrilled by our life choices. I have a feeling he would have visited us in Africa many, many times. I have often felt his presence throughout these years in Kenya. I am certain he is here with us, guiding us in this adventure.

If I have shared my story, it is also thanks to him. To his immense love for nature, for animals, and for his dear Africa. To his infinite love for Luca. It is thanks to his ability to pass down his deep passion for Africa to his son that I find myself here today, in Kenya, looking out at Kilimanjaro, with a pen in hand, with so much to remember and to write...

Almost 30 years after my arrival in Africa, I stop to look within myself, aware of the changes in me, of my evolution. Aware of the fundamental role Africa has played in this "becoming."

Almost 30 years have passed since I arrived in Kenya, and I am still here, contemplating the same plains scattered with proud animals, with the majestic Kilimanjaro in the background, still incredulous at such beauty, such tranquility, such magic.

These have been years full of passion, love, and adventure; sometimes, there were also difficulties, disappointments, and discouragement. But they were years' worth living.

I am more of a woman, more mature, more aware. During these years in Africa, there were times when I lost myself, but never for too long. I always found myself again, and I am certain that living in Africa has contributed to whom I have become. I have had moments of doubt, fear, and uncertainty, but then peace always returned to my heart, and my soul found harmony again.

My children are growing up, and Luca and I are adapting, organically and naturally, to their needs and desires.

Lucrezia is 22 years old; she attends the University of St. Andrews in Scotland, finding success and satisfaction. She studies Sustainable Developments and Geography. She is now a young woman, passionate, sweet, cheerful, intelligent, compassionate, and generous. Her soul is full of light, and it is a joy to have her by my side. She teaches me something every day. She is a daughter of Africa and will always be.

Jacopo and Lorenzo are growing up a little wild but are truly adorable! Their hearts are pure and full of beautiful energy. Both are studying at a boarding school in Switzerland and come home to Kenya for their holidays four times a year. Of course, I miss them terribly, but I know that this life experience is crucial and formative for their character and personality. They are going through the challenging teen years, and it is not always easy to be with them. I like to watch them in silence, treading lightly, and I can already catch a glimpse of the strong, honest, and courageous men they will become.

More than 30 years after our first meeting in Brescia, Luca is still by my side. Certainly, sometimes, being together for so many years has not always been easy. But with time, when you live side by side, you learn to see things with four eyes instead of two, and to feel and love with two hearts, in unison. You learn to sing a single song in harmony, though with two voices. And this makes your life more special than ever.

We still hold hands when we walk at sunset. We support and help each other in difficult times. We deeply respect each other and hold the same values. The truth is that our hearts still beat fast when we see each other after being apart for a few days, and we are friends before lovers. Perhaps this is the secret of still being together after so many years.

During these years, I have traveled a lot. I have realized that sometimes I need to go to new places, to leave the savannah for a while, to search for what I have not yet found. Only to desire to return here.

Sometimes, I travel with Luca for work and other reasons; these escapes from our daily life are precious opportunities to be together, just the two of us, once again being romantic and feeling in love, just like at the beginning of our relationship.

Sometimes, I travel alone for pleasure. I often go to Nairobi for a few days to immerse myself in its chaos, see friends, and shop; sometimes, I go to Malindi to fill my eyes and heart with the infinite spaces and magical colors of the ocean.

I often go to Italy to see my family and lifelong friends, to embrace those who still love me so much and miss me every day.

But it is only when I return to Kenya, to the green Chyulu Hills, among the Maasai and immersed in wild nature, that my heart calms, my anxiety subsides, and I reconnect with myself and my soul.

Campi ya Kanzi is the only place where I feel at home, where I am at peace with myself, where I accept and love myself for who I am, and where I find harmony and light within me. The only place where I find the true Antonella, the true Naseryan.

In recent years, our African family has grown. Ten years ago, Mia, a spirited and intelligent Jack Russell Terrier, entered our lives, filling our days with joy and companionship. She is, quite literally, my shadow – protecting me with fierce loyalty and keeping me company in my African days. Four years ago, Akila arrived, a magnificent Rhodesian Ridgeback, strong and majestic, tender and affectionate. He has become Luca's shadow, following him everywhere, always by his side. These dear friends live in harmony with us, giving us moments of pure love that are precious and unforgettable.

A special thank you goes to Luca, who has always believed in me and convinced me to turn this diary into a real book. He has read and reread my stories with patience and dedication, serving as my first real editor! And to my children, who give meaning to my existence and light up my life. My daughter Lucrezia graciously translated my words into English, and I will always be grateful to her for this.

I would like to also thank our families in Italy and Africa and all the dear friends, named and unnamed, whose lives have inspired my writings and enriched my existence and me as a person.

My beloved Zia Renata and Zia Gabriella are sadly no longer with me. I feel their absence every day. A dear friend that I have mentioned in my book has, unfortunately, left us. He is in another dimension, and we still feel him close to us. I know we will meet them again in a better place. Other friends have left Kenya forever; other couples are no longer together because their lives have taken different directions. But our feelings for them, wherever they are now, have not changed, and their memory is always with us.

Every morning, when I wake up, I smile and feel deeply grateful for what I have: an extraordinary life, full of light and good energy, which I feel obliged to live with awareness and wonder every day.

What was meant to be, is. I do not know what tomorrow holds; life is a wonderful journey, an adventure that renews itself every day, filled with new lessons. With so much to see if you keep a watchful gaze, and to feel if you have an open heart, eager to love. I am ready for whatever life has in store for me. And I do not feel alone. Time passes, and life is in constant flux. People change, situations do as well, and I try to adapt to change with constant enthusiasm and curiosity.

For me, the beauty of living in Africa is understanding how to help those who truly need it; it is the awareness of being here, on this Earth, with a specific purpose. Giving and not just taking. It is feeling like I am part of a whole, always and in every way. It is the wonder and enchantment before the marvel of nature every day. It is the kind and wonderful hearts of its people, who, even in the face of poverty, misery, and misfortune, do not give up and continue to live without losing hope and faith in their destiny. Always.

This is the greatest lesson that Africa and its people have taught me, a lesson I pass on to my children each day.

I would like to offer a special thank you to Titus and Samson, who have shared our lives since the beginning of our African adventure, always helping and supporting us.

Titus Muia is a wise, intelligent man, a valuable advisor, and a capable, honest director of our foundation (The Maasai Wilderness Conservation Trust). He is always eager to embrace our new ideas and bring them to life, working with us for the good of the Maasai and the land they inhabit.

Samson Parashina, the son of a great and respected Maasai chief, began working at Campi ya Kanzi 25 years ago, cutting grass around the camp. He became our best waiter and, finally, an excellent professional guide. For several years, he was the manager of Campi ya Kanzi and is now the chairman of our foundation (The Maasai Wilderness Conservation Trust). A determined, intelligent, and curious man, he has made the preservation of his land and his Maasai people the purpose of his life. His three sons – Sarijore, Sankale, and Sankei – are the same age as my children. They have grown up together, respecting and loving each other as siblings. This brings me great joy and fills our lives with meaning. Perhaps one day, mine and Samson's children will work together to preserve this land, its inhabitants, and the animals that call it home.

Finally, I want to thank the entire staff of Campi ya Kanzi and the Maasai Wilderness Conservation Trust, as well as all the Maasai of the Kuku Group Ranch, for believing in our dreams and helping us make them a reality.

What we achieve each day is part of a grand adventure, and I am proud to have people beside me who believe in us and who are dedicated to preserving our beloved Africa.

Antonella Bonomi
Campi ya Kanzi, May 2025.

Photographs

Mum and Dad

Little Antonella in Forte dei Marmi in 1970

Antonella and her sister Chiara

Luca and his mum in the Maasai village of Iltilal in 1975. This is the very same village nearby the site where Campi ya kanzi will be built 20 years later

The beginning of Campi ya kanzi: the construction of the kitchen

A young Luca adorned with the Maasai warrior decoration

Antonella and Moci with Lupembe at the Maasai village

Antonella and Moci having tea with the Maasai women

Luca jumping with Daudi and Munghesia

Our first camp while we were building Campi ya Kanzi

Our little tent at the beginning of our adventure in Kenya

The mess tent where we used to have our meals

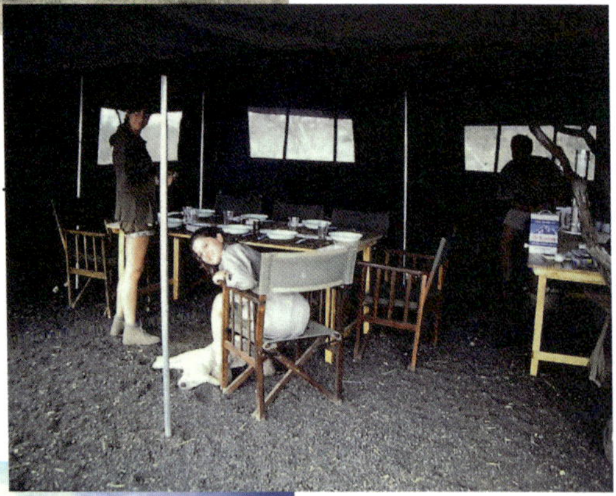

The building of Tembo House, the main area of Campi ya Kanzi

Antonella and Luca's wedding

Antonella in her wedding gown

Lucrezia and Katito

Luca and Lucrezia in the Chyulus clouds' forest

Lucrezia and the Maasai warriors

Lucrezia and the Maasai women

Lucrezia with Jago, Kibo and Womble

Luca and our beloved Jago

Patrick, our chef 1998

Lucrezia dancing with the Maasai

Lucrezia at Longido with some Maasai

Lucrezia and her maasai friends of Kanzi Academy

Lucrezia and Sarijori, the son of Samson

Lucrezia and Nashipei

Lucrezia and Katito

Lucrezia on safari with mum and dad

Luca and Antonella on safari

Us with Kibo

Antonella with her cubs (credits to Nicola Tonolini)

My sweet ballerina

Lucrezia and Pongo

Campi ya Kanzi staff

Matasha, Antonella and Luca in Peponi walk

Antonella, Luca and Jago in their tent

An happy Antonella with Maasai children

Anna, Clare, Lucrezia and the students of Kanzi Academy

Luca relaxing in the forest

Luca and Jacopo walking in the savana

The Belpietros at home (credit Benedetta Mazzini)

Camping time in savana

Antonella with elephants

Antonella, Lucrezia e Luca walking in the savana (credit Benedetta Mazzini)

Lorenzo and his Maasai friends at the Kanzi Academy

Sarah, Jacopo and his maasai friends at Kanzi Academy

Lorenzo and Maasai children

On safari with our Land Rover

Luca with Jacopo and Lorenzo

Lorenzo and Matasha in the forest

Lorenzo learning to fly next to Kilimanjaro

Lucrezia outside our home in Ponte di Legno in Italy

The Malindi house

The Castellino in Italy

The beautiful beach of Kiwayu

Antonella and Lucrezia walking on the beach of Kiwayu

Lucrezia enjoying the view of the bay of Kiwayu

Mia and Akila

Luca and Julian

Nkalami walking in the savana

Happiness in the bush

Peponi Walk and the forest on the top of the Chyulu Hills

The view from Okoikuma

The majestic Kilimanjaro

Index

Freddie del Curatolo

journalist and writer

Africa's is an infinite song sung by millions of voices, human and animal, that has always sought its harmony and often, extraordinarily, manages to find it.

Antonella Bonomi, unlike Karen Blixen who was the first to transcribe the score, does not ask herself if Africa "knows its song", because she herself has become part of those melodies.

Narrating what is a "choice of love", with a captivating style between autobiography and novel, pages recovered from a diary saved from the fires and from the surprises of life in Kenya alternate with frescoes of wild Nature, respect and deep gratitude for a Land that "can give a lot, but can take away just as much, suddenly and without a reason".

"Canto d'Africa" does not celebrate the entrepreneur who with her husband Luca Belpietro created and manages Campi Ya Kanzi, one of the most magical, sustainable and renowned realities for safaris in the savannah, but celebrates the deep gratitude to the African continent and its life lessons.

From the Chyulu Hills, which are not only the "green hills of Africa" at the foot of Kilimanjaro described by Ernest Hemingway, we travel through the wonder and variety of landscapes and situations of Kenya and Tanzania, from the nature reserves of Maasai Mara and Samburu, to the lunar Lake Turkana, from modern Nairobi with its contrasts and colonial reminiscences to the remote dream island of Kiwayu, from the majestic crater of Ngorongoro to the quiet of Malindi, where the best friends of the houses are the gardens.

The journey of a passionate and courageous woman, of a proud and present wife and mother, a life story that, also thanks to the wonderful setting in which it is set and the author's ability to narrate it, enriches us and does us good. Singing about that Africa that teaches "every day a little to live, and a little to die".

RICCARDO ORIZIO

journalist and writer

Thirty-five years after "I Dreamed of Africa" by Kuki Gallmann, another extraordinary, moving and even more unforgettable story of real life emerges from the savannahs of Kenya. Antonella Bonomi's is a personal, almost intimate diary, written primarily for herself and her family, a collection of true adventures, including those of the soul and the heart, that unite the Africa of yesterday to that of today. "Canto d'Africa" is not a reportage of the continent's miseries, nor is it the artificial exotic dream of an easily impressionable tourist in search of escapism. It is a one-off, impossible to duplicate, true journey of a woman who life (and the ambitious project of a husband who is pioneer of nature conservation) takes from a bourgeois and urban life to the uncontaminated wilderness of Africa among the maasai warriors. Here, at the foot of Kilimanjaro, little by little she discovers a vocation, for herself and for her family, that goes beyond the postcard image of her company's luxury safaris and becomes a brave human adventure that, before the last page is reached, each of us would have dreamed to experience.

"Antonella Belpietro's lyrical account of life among the Maasai in southern Kenya is Italy's answer to Karen Blixen. Read it and be bewitched".

Brian Jackman

Antonella Bonomi

Antonella Bonomi is a pioneer and entrepreneur who has dedicated her life to blending luxury hospitality with sustainability in one of the world's most iconic wilderness areas, the Chyulu Hills, in Kenya.

Born and raised in Italy, Antonella studied law and briefly practiced before immersing herself in her first passion: the family wine business. Her early career in the wine industry instilled in her a deep appreciation for heritage, craftsmanship, and the natural environment—values that would later define her life's work.

Falling in love with Luca, her husband with whom she has three children, meant falling in love with Kenya, where they moved together over 30 years ago. There she embarked on an extraordinary journey to create a harmonious balance between tourism, Maasai culture, and environmental preservation. Together, Luca and Antonella co-founded **Campi ya Kanzi and Maasai Wilderness Conservation Trust**. Kanzi is one of Africa's first eco-lodges powered entirely by renewable energy and recognized globally for its regenerative tourism model. Nestled in the shadow of Mount Kilimanjaro, the lodge represents Antonella's unwavering commitment to protecting the environment while uplifting the local Maasai community.

Antonella's journey has been one of bold choices and transformative experiences, from navigating the challenges of building a lodge in the remote Chyulu Hills to raising three children in the bush and forming lifelong friendships with the Maasai.

Antonella's memoirs weave together her personal story with the beauty, complexity, and spiritual depth of life in the African wilderness, reflecting the transformative journey shaped by her bold choices. Through her words, she shares the triumphs and challenges of building a life rooted in purpose, love, and legacy.

When not managing her lodges, Antonella enjoys exploring new destinations and spending time with her family. Her memoirs are a testament to the power of love, dreams, determination, and the enduring connection between people and the land.

Printed in Dunstable, United Kingdom

74170150R00161